PRACTICES OF THE SELF AND SPIRITUAL PRACTICES

Practices of the Self and Spiritual Practices

Michel Foucault and the Eastern Christian Discourse

Sergey S. Horujy

Edited with an introduction by
Kristina Stoeckl

Translated by
Boris Jakim

WILLIAM B. EERDMANS PUBLISHING COMPANY
GRAND RAPIDS, MICHIGAN / CAMBRIDGE, U.K.

First published in Russian under the title
Poslednii Proekt Fuko. Praktiki Sebja i Dukhovnie Praktiki.
In *Fonar' Diogena. Kriticheskaya Retrospektiva Evropejskoi Antropologii,*
© 2010 Institut Filosofii, Teologii i Istorii sv. Fomy, Moscow.

English translation © 2015 William B. Eerdmans Publishing Company
Grand Rapids, Michigan / Cambridge, U.K.
All rights reserved

Wm. B. Eerdmans Publishing Co.
2140 Oak Industrial Drive N.E., Grand Rapids, Michigan 49505 /
P.O. Box 163, Cambridge CB3 9PU U.K.

Library of Congress Cataloging-in-Publication Data

Khoruzhii, S. S. (Sergei Sergeevich)
[Poslednii Proekt Fuko. Praktiki Sebja i Dukhovnie Praktiki. English]
Practices of the self and spiritual practices: Michel Foucault and the Eastern Christian discourse /
Sergey S. Horujy; edited with an introduction by Kristina Stoeckl; translated by Boris Jakim.
pages   cm
Includes bibliographical references and index.
ISBN 978-0-8028-7226-5 (pbk.: alk. paper)
1. Foucault, Michel, 1926-1984. 2. Philosophical anthropology.
3. Theological anthropology — Christianity. 4. Synergetics. 5. Asceticism.
I. Stoeckl, Kristina, editor, writer of introduction.   II. Title.

B2430.F724K4613   2014
194 — dc23

2014041217

www.eerdmans.com

Contents

Editor's Introduction ... vi
Preface to the American Edition ... xvi

Introduction ... xxiii

I. Foucault's Last Project, or: Hermeneutics by No Means of the Subject ... 1
   I.1. The Language of Foucault's New Conception of the Subject ... 5
   I.2. Sketch of the Conception ... 26
       I.2.1. Genesis of the Practices of the Self: The Platonic Model ... 27
       I.2.2. The Hellenistic, or "Ethical," Model ... 30
       I.2.3. The Christian, or "Religious," Model ... 46
   I.3. Outline of Foucault's "General Project" and Discussion of His New Conception of the Subject ... 65

II. Spiritual Practice, Synergic Anthropology, and Foucault's Project ... 99
   II.1. Reconstruction of the Hesychast Practice ... 102
   II.2. From Hesychasm to Synergic Anthropology ... 123
   II.3. Historical Sequence of Anthropological Formations ... 131
   II.4. Anthropological Scenarios and Projects for Modernity ... 143

III. So Where Shall We Sail? ... 165
   Bibliography ... 176
   Index ... 179

Editor's Introduction

Sergey Sergeevich Horujy is one of the most prolific philosophers in Russia today. Born in 1941, he studied physics at Moscow State University, received his PhD in physico-mathematical sciences in 1977, and served as professor of mathematical physics at the Steklov Mathematical Institute of the Russian Academy of Sciences until 2006. He has published numerous articles and books in this discipline, both in Russian and in English. But besides his career in the natural sciences, an unsuspected profession under the antireligious Soviet Communist regime, Horujy was a clandestine, tireless student of Russian religious philosophy and Orthodox theology. He became a member of Moscow's underground religious *intelligentsia* of the 1970s and 80s and wrote a great number of philosophical works, all of which could appear in print only after 1991. To add to this already-impressive body of scholarship, he also engaged in literary criticism and translation, becoming the translator of James Joyce into Russian. Since the end of communism, Horujy has emerged as the leading figure of the renewal of religious philosophy in post-Soviet Russia, to which he has given an original "anthropological" twist. Today, he is professor of philosophy at the Institute of Philosophy of the Russian Academy of Sciences and founding director of the Institute of Synergic Anthropology.[1] He travels and lectures frequently in Russia and abroad, and several of his philosophical essays have been translated into English.[2] This book is the first

---

1. I refer the interested reader to the website of this institute. The website includes an extensive library-section with direct access to most of Horujy's works in Russian and English: http://synergia-isa.ru/.

2. Some works by Sergey S. Horujy published in English translation: "Globalistics and

Editor's Introduction

book-length translation of a work by Horujy into English and represents the most current stage in his long and impressive course as a philosopher. In order to help the unacquainted reader to enter into Horujy's philosophical universe and terminology, this introduction tries to give an overview of the development of his thought: from his beginnings as a student of the religious philosophy of the pre-revolutionary years up to his engagement with postmodern French philosophy and the project of synergic anthropology.

Horujy's interests range from pre-revolutionary Russian religious philosophy to the theology of the Church Fathers to modern and postmodern Western philosophy. The works in which he covers this wide range of topics were published in brief succession from 1991 onwards. They testify to decades of intellectual engagement during which Horujy developed his own theological and philosophical position. During one of our conversations Horujy remembered that during his student years, the first food for thought for anyone interested in philosophy and religion was, by default, the works of the philosophers of the Silver Age. Consequently he wrote extensively about pre-revolutionary Russian religious philosophers, notably about Alexey Homyakov, Vladimir Solov'ev, Pavel Florenskij, Sergei Bulgakov, and Lev Karsavin.[3] Horujy's name in the West is firmly connected with these studies and he continues to be invited to conferences on pre-revolutionary Russian religious philosophy, even though his own original philosophical work has decidedly moved on from these beginnings. Actually, he himself describes his philosophical career as a moving away from the "methodological sloppiness" of the pre-revolutionary religious philos-

---

Anthropology: An Approach to the Problem," *World Public Forum Dialogue of Civilizations Bulletin* 3 (2004); "The Idea of Total Unity from Heraclitus to Losev," *Russian Studies in Philosophy* 35, no. 1 (1996); "Philosophy vs. Theology: New and Old Patterns of an Ancient Love-Hate," *Philotheos* 1 (2001); "*Ulysses* in a Russian Looking Glass," *Joyce Studies Annual* (1998); "Vladimir Solov'ev's Legacy after a Hundred Years," *Russian Studies in Philosophy* 46, no. 1 (2007); "Breaks and Links: Prospects for Russian Religious Philosophy Today," *Studies in East European Thought* 47, no. 1-2 (2001); "Man's Three Far-Away Kingdoms: Ascetic Experience as a Ground for a New Anthropology," *Philotheos* 3 (2003); "*The Brothers Karamazov* in the Prism of Hesychast Anthropology," *Sofia Philosophical Review* 3, no. 1 (2008); "Crisis of Classical European Ethics in the Prism of Anthropology," Preprints of *ISA* 3 (2007); "Personalistic Dimensions of Neo-Patristic Synthesis and Modern Search for New Subjectivities," *Theologia* 81, no. 4 (2010); "What Is *Synergia*? The Paradigm of Synergy in Its Principal Subject Fields and Discursive Links," Preprints of *ISA* 5 (2011). These and other translated essays and speeches can be downloaded from the website; see footnote 1.

3. Monographs in Russian: *Mirosozercanie Florenskogo* (1999), *O starom i novom* (2000), *Opyty iz russkoi duhovnoi traditsii* (2005), *Posle pereryva. Puti russkoi filosofii* (1994).

ophers to the theological rigor of the neo-Patristic theologians, which he then sought to translate into his personal philosophical language of synergic anthropology.[4]

It is probably no exaggeration to say that the decisive element of renewal and originality in Horujy's program of Russian religious philosophy is connected with the neo-Palamist Patristic theology of the Russian emigration, notably with the works of John Meyendorff. During the first interview I held with Horujy in 2005, he recalled how, in the early 1970s, he first found out about new developments in Russian émigré-theology. In those years he came across John Meyendorff's doctoral thesis on the late-Byzantine Church Father Gregorios Palamas. The book was in French and had somehow passed the censorship for religious literature unnoticed. Unlike other literature of the genre it was not kept in the reserved sections of the *spec-hran* (from *special'noe hranenie*, "special storage") of the university's library, but it could simply be ordered from the librarian's desk. Horujy said that he immediately felt intrigued by this "new way of philosophical and theological reflection." He consequently concentrated his studies on the Church Fathers and on the history and theology of Hesychasm. Under a pseudonym he translated theological literature into Russian. There is anecdotal evidence for the complete secrecy of this translation-work: not long ago Horujy found himself invited as a guest-speaker to the book-launch of a re-edition of the Russian translation of "Einführung in das Christentum" ("Introduction to Christianity") by Joseph Ratzinger (the future Pope Benedict XVI), the very same translation which he, unbeknown to the new publishers, had made himself many years ago. Today, Horujy is considered a major authority on Hesychasm. He is a member of the Biblical-Theological Commission of the Patriarchate of Moscow and is frequently invited to theological conferences in Russia and in the West.

Horujy's critical distancing from the canons of Russian religious philosophy was to a considerable degree induced by his discovery of neo-Patristic theology. A long essay by him about the émigré philosophy and theology of the 1920s and 30s bears the title "A Step Ahead, Taken in Dispersal."[5] In this text he makes it clear that for him the main intellectual achievement of the Russian diaspora was made in the field of theology.

---

4. In early English translations, also by the author himself, one finds the expression "synergetic anthropology." From 2010 onwards, the term "synergic anthropology," a more correct translation from the Russian "sinergijnij," is used.

5. "Shag vpered, sdelannyi v rasseianii," in *Opyty iz russkoj duhovnoj tradicii* (Moscow: Izd. Parad, 2005).

Editor's Introduction

Horujy credits the neo-Patristic theologians with having changed the orientation of Orthodox thought. The philosophers of the Russian "Silver Age" in his view no longer had much to offer to contemporary philosophy, and Horujy is therefore dismissive of the attempts at their revival made by some of his contemporaries.

The encounter with neo-Palamism greatly helped Horujy's reflection on the necessity of a new discourse different from that of Russian religious philosophy, taking more rigorously and correctly into account what he called later the Eastern Christian discourse, both the dogmatic/theological and ascetical/practical experience of Eastern Christianity. A large part of Horujy's work is dedicated to the notion of hesychasm and asceticism. The erudite response that he gives to Foucault in this book, respectfully and firmly pointing out the inadequacies of Foucault's reconstruction of a "Christian model" of practices of the self, is based on his own accurate study of the phenomenon of Christian spiritual practice and the hesychast tradition.[6] The neo-Palamist theologians effected a turn for Orthodox thought that Horujy, with reference to Heidegger, calls *Kehre*: a (re-)turn, or a "modulation of the discourse."[7] However, this turn was a theological, not a philosophical phenomenon, Horujy writes, and when this thought could finally make its way back to Russia after the fall of communism, its philosophical potential had not yet been explored. To do this is exactly the task Horujy sets himself: "Russian philosophy stands in front of a new beginning," he writes — a new beginning that implies a rethinking of the relationship between theology and philosophy as it has been manifest in classical metaphysics.[8] The elaboration of an anthropology that would overcome the limitations imposed by Western metaphysics is Horujy's philosophical project. This is precisely how we have to understand the following passage in this book, a passage that must absolutely not be overlooked because it represents a cardinal statement of intent, for this book and for Horujy's philosophical oeuvre in general:

> After the complex peripeteias of the development of Russian thought in the twentieth century, characterized by a tangled relationship between its European and Eastern Christian contexts as well as between its philosophical and theological discourses, and given the restoration,

---

6. *Sinergija: problemy asketiki i mistiki pravoslavija* (1995), *K fenomenologii askezy* (1998), *Hesychasm: An Annotated Bibliography* (2004); *Issledovaniya po isihastskoi traditsii*, vols. 1 and 2 (2012).
7. Horujy, *Opyty*, p. 28.
8. Horujy, *Opyty*, p. 29.

EDITOR'S INTRODUCTION

"after the interruption," of the possibility of free philosophizing in Russia, there arose the undoubted necessity of *"another beginning,"* of a *new* reflection of philosophical thought on its *own* sources and foundations, on its own "two-context" (European and Eastern Christian) nature. (on page 101 in this book)

Horujy's starting point, not only in this book but in his entire philosophical project, is what he perceives as a crisis of humankind. This crisis, he writes repeatedly, is not only a Russian, but a global phenomenon, even though it might have found a particularly sharp expression in post-Soviet Russian society where many people suffered an existential loss of orientation in life with the collapse of the old regime. This crisis, which has also been described as the "crisis of the European subject,"[9] is of an anthropological nature for him; it has to do with the way in which human beings conceive of themselves, how they take a place in the world and vis-à-vis each other. It is a crisis of modern philosophy, where the human being was conceptualized in terms of subject, substance, essence. Horujy attributes the formulation and perfection of this understanding of man, the "classical European anthropological model," to the intellectual legacy of Aristotle, Boethius, and Descartes:

> For a long time, a model [of the human subject] dominated European thought in which . . . the identity of a person was understood . . . as founded on substantiality. In the classical European anthropological model, human nature bore the character of a substance: Completing the anthropology of Aristotle, which understood man as a definite system of substances, Boethius, at the beginning of the sixth century,[10]

---

9. The phrase "crisis of the European subject" refers not only to the book by Julia Kristeva that carries this title (Julia Kristeva, *Crisis of the European Subject* [New York: Other Press, 2000]) but to a more general topic in contemporary philosophy, namely the deconstruction of the human subject in philosophy. "Who comes after the subject?" is a question posed in an edited volume of French philosophy (Eduardo Cadava, Peter Connor, and Jean-Luc Nancy, eds., *Who Comes after the Subject?* [New York, London: Routledge, 1991]). Horujy's work, which repeatedly makes reference to this book, must be read as one attempt to answer that very question.

10. Horujy is referring to the late-Roman philosopher and statesman Anicius Manlius Severinus Boethius (480-524), whose translations of Aristotelian and Platonic philosophy into Latin had a decisive influence on scholasticism and Western philosophy. Especially the translation of Aristotelian terminology in logic is of relevance here, for example the Greek *ousia* into Latin *substantia*.

advanced the famous definition according to which man is an "individual substance of rational nature." Later on the concept of subject (a thinking subject, the subject of reason) was added to this definition, and from here emerged the perfect construction of man in its impenetrable philosophical armour: the classical European man of Aristotle, Boethius and Descartes as an essence, a substance and a subject. And as self-identity.[11]

In the book *Diogenes' Lantern*, the last chapter of which is here presented as a short monograph, the argument about the birth of the classical metaphysical subject out of the Latin Christian appropriation of Greek philosophy and its consequent development in Western philosophy is laid out in all its detail. The first part of *Diogenes' Lantern* sketches out, in chronological order, the conceptualization of the human subject from Aristotle to Boethius to Descartes to Kant and all the way up to German idealism. This exposition of the classical European anthropological model occupies roughly 200 pages (one third) of the book. The relatively brief treatment of the classical metaphysical subject in this book is indicative of the fact that Horujy does not linger on the problematic because he considers it a development that is, in some sense, over and done with. Here he differs from "civilizational" advocates of Orthodox thought like Christos Yannaras or Dumitru Staniloae, who maintain that the West is forever bound to its classical metaphysical foundations, using this argument as a proof for culturally and historically grounded differences between the East and the West.[12] What is important for Horujy is the fact that this classical human subject, man as an essence and a substance, has increasingly been put into question since the late nineteenth century. The crisis of modern times lies precisely in the becoming-unfounded of the Aristotelian-Boethian-Cartesian subject, and Horujy reads Western philosophy in the twentieth century as a document of this crisis, a crisis that was expressed in Nietzsche's critique of Enlightenment rationality and subjectivity and in Heidegger's dismantling of classical metaphysics. Against this background Horujy turns to contemporary philosophy, and singles out two philosophers who have responded to this crisis by trying to think "beyond the subject": Kierkegaard and Foucault.

11. Horujy, *Ocherki sinergijnoj antropologii* (Essays in Synergic Anthropology) (Moscow: Institut filosofii, teologii i istorii Sv. Fomy, 2005), pp. 78-79.

12. Mihail Neamtu, "Between the Gospel and the Nation: Dumitru Staniloae's Ethno-Theology," *Archaeus* 10, no. 3 (2006); Christos Yannaras, *Orthodoxy and the West: Hellenic Self-Identity in the Modern Age* (Brookline, MA: Holy Cross Orthodox Press, 2007).

He situates his own contribution, his "new anthropology," in exactly this philosophical realm.

The present book singles out the chapter of Foucault in the book *Diogenes' Lantern*. This chapter is suitable for a free-standing publication because of its clear focus (the dialogue between the author Horujy and Foucault) and accessibility to a Western readership (it starts off with a well-known Western author and reaches out to the less well-known Eastern Orthodox tradition). Horujy's interpretation of Foucault is highly original and up to date, and goes against the grain of the mainstream of Foucault-scholarship. Focusing on the last four years of the French philosopher's life, Horujy identifies in the last works of Foucault a "new" type of philosophical project and alternative to the classical European anthropological model. His careful reading of *The Hermeneutics of the Subject* works out the novelty of this book: the shift of the philosopher's focus from the practices of power to the "practices of the self."

Horujy is an expert on Christian ascesis, and it is therefore an easy task for him to pinpoint the shortcomings in Foucault's reconstruction of the Christian ascetic model. For Foucault, Christian "practices of the self" could not offer a viable alternative, a fruitful way out of the "crisis of the European subject." His preference clearly lay with his reimagined Hellenistic model of "esthetics of existence." However, Horujy's reading of *The Hermeneutics of the Subject* makes clear that Foucault's reconstruction of the Christian "practices of the self" was inconclusive not only because the author may have lacked historical theological expertise, but because, as Horujy writes on page 157, "there is confusion between the spirituality of the Christian West and that of the Christian East." It is in the Christian East, in particular in the spiritual tradition of hesychasm, that Horujy identifies a viable model of "practices of the self" and the ground for developing an anthropological alternative to the classical European subject. This is what Horujy says about Foucault's understanding of the Christian "practices of the self" on page 57:

> Foucault has a very good sense of the embryonic movements of thought in a monk's consciousness, movements that Western philosophy, as a rule, does not feel or know; but Foucault does not grasp what exactly it is that the ascetic consciousness does with these movements.

However, whether Foucault "does not grasp" what exactly Christian ascesis is about because the only Christian model accessible to him was Catholicism, or because he was steeped in the tradition of French anticlericalism (both hypotheses sustained by Horujy), is, in a way, secondary. What is important — and the originality and novelty of Horujy's interpretation lies precisely in individuating this fact — is that Foucault's later philosophy prepares the ground for a new kind of philosophical anthropology, an understanding of the human subject no longer in terms of essence or substance, but in terms of "practices."

It is precisely this ground that Horujy, in the second part of the book, occupies with his own philosophical project and anthropological "proposal": *synergic anthropology*. In this second section, Horujy turns from the dialogue with a work of contemporary philosophy to present his own, independent work of contemporary philosophy, rooted in but not limited to the Eastern Christian philosophical and theological tradition. What does this project consist in? At the heart of synergic anthropology's attempt to offer an alternative to the Cartesian subject lies the realization that the Orthodox tradition is built around an experience that Cartesian metaphysics cannot account for: the experience of *theosis*, deification. This experience is described in the ascetic literature of the Fathers of the Desert and it is explained in the theology of hesychasm. Its basic element is the understanding that man exists vis-à-vis another form of being and that a transformation of human being in view of this "Other-being" is possible. Horujy reminds the reader that once we take the anthropological reality of mystical experiences and spiritual practices seriously, we are inevitably led to a reconsideration of the classical anthropological paradigm of man as an autonomous, self-centered subject. Where before we would have man as an essence and a center, and where the postmetaphysical philosophers of the twentieth century identified a lack, Horujy puts man as an energetic constellation and a pluralistic being endowed with a triple-border. The main point is that these borders are not closed, but that they are realms in which processes of interaction with "Other-being" can take place. These processes aim at what Horujy calls "unlocking" *(razmykanie)*, the interaction of man's manifestations with the energies of the "Other." From an "anthropology of the border," Horujy has thus moved on to an "anthropology of unlocking," *synergic anthropology*.

At first reading the project of synergic anthropology may seem slightly eccentric, and at times Horujy's highly concentrated, almost technical language makes for challenging reading. What Horujy offers is a philosophical anthropology that is open — but not limited — to the reality of ascetic-mystical experiences, a philosophical anthropology that has a place for religious experience while not being a religious anthropology itself. This is maybe the most important, but also the most difficult point in Horujy's entire project: synergic anthropology develops on the grounds of a reflection on the Eastern Christian hesychast tradition, but at a certain point (in the reflection on the "ontic" and "virtual" border) it leaves this tradition behind and becomes a universal anthropological model. The philosophical project itself does not prejudice any particular kind of strategy of self-realization:

> [S]ynergic anthropology . . . does nothing more than affirm that Man has initially a certain inner pluralistic character, implanted in the very structure of his constitutive, extreme experience. This affirmation does not by any means imply that an individual must embody and cultivate this pluralistic character in his strategies . . . synergic anthropology does not yet decide what an individual's attitude should be toward this trait of his (it sees here a special problem, requiring a very different context for its discussion). (page 132)

This point, the principled openness of synergic anthropology, may become clearer if we read Horujy's "new anthropology" against the background of another "new" take on ontology in the twentieth century. Martin Heidegger called the forgottenness of Being *(Seinsvergessenheit)* the major shortcoming of classical metaphysics. His *Fundamentalontologie* was designed as a response, but we know that Heidegger himself did not escape, in 1934 and however briefly, the temptation of regrounding this "new" ontology in an essentialism of the most destructive kind. I would like to suggest that Horujy's synergic anthropology can be read both as a commentary on Heidegger's failure and as a response to Heidegger's question. It is a commentary on the failure of totalizing a discourse inasmuch as it lays out an anthropological model in which the question of Being can be asked once again, but in which it is not asked exclusively. Horujy conceptualizes man in pluralistic terms, as determined by a triple border, not in terms of the ontological border only. His synergic anthropology is a response to the Heideggerian problematic of de-essentializing the human subject because it looks at the person not in terms of essence, but in terms of manifestations, as energy and potentiality vis-à-vis an "Other." It is a response that draws on Orthodox theology; and, far from the culturalist fervor of so many interpreters of Orthodox theology, Horujy invokes this intellectual tradition in the engagement with a problematic that cannot be limited to the West or to the East.[13]

13. I have explored in depth this important "post-totalitarian" character of Horujy's

Editor's Introduction

In the preface to this edition, Horujy writes that his philosophical method to explore and make fruitful the improbable dialogical relationship between the Eastern Christian spiritual tradition and modern Western philosophy reflects the essentially postsecular nature of our contemporary period. I agree with him and would like to make this point even stronger: Horujy's thinking not only exemplifies the postsecular nature of our contemporary period, it actually represents the idea of postsecular dialogue at its best; it is a powerful invitation to think *beyond: beyond* religious-secular divides in philosophy, *beyond* East-West divides in intellectual history, *beyond* the subject and *towards* the human being.

The idea for the translation of this book has come out of several years of scholarly collaboration with S. S. Horujy, during which I felt a growing sense of urgency to render his philosophical work accessible to a non-Russian readership. This book can only be a small step in this direction. As editor, I would like to thank the following persons and institutions for their support of this project: the Gerald Palmer Eling Trust and the Austrian Academy of Sciences for funding the translation, Boris Jakim for his excellent translation, and William Eerdmans for taking an interest in this project.

KRISTINA STOECKL
Vienna

---

philosophy in my book *Community after Totalitarianism: The Russian Orthodox Intellectual Tradition and the Philosophical Discourse of Political Modernity* (Frankfurt: Peter Lang, 2008).

Preface to the American Edition

There is but one Big Theme of modern reflection on man and his situation: profound *tectonic shifts* that take place in the human being, in structures of his personality and his relation to himself, in his position and strategies in social, technological, environmental reality, and so on. These shifts represent a strong challenge to this reflection: they generate the crisis of old anthropological theories and imply the necessity to create new ones. The whole fund of old anthropological views, ideas, and conceptions needs global revision and reappraisal. One should answer the question: How should man apprehend his present self and his situation, which has changed radically and continues to change? How should he act and how can he achieve his full self-realization in this situation?

I have always considered my work as an ordered (as far as possible) series of studies and sketches on this Big Theme.

As for the present small book, it must be said first of all that it is a part of a much bigger book: it is the concluding section of my Russian monograph *Diogenes' Lantern: Critical Retrospect of European Anthropology*. "Anthropology" is conceived here as philosophical anthropology and the monograph presents a critical retrospection of the evolution of philosophical thought on man, from the first conceptualization of man in Aristotle to the theory of practices of the self by Michel Foucault. In the prism of the present-day anthropological situation, the way of this anthropology looks correlative to the way of philosophy according to Heidegger: the "forgetting of being" finds its correlate in the "forgetting of man." After touching briefly upon initial stages of European anthropology, the book concentrates on disclosing the genesis of what I call the "anti-anthropologism" of classical

Preface to the American Edition

metaphysics in Descartes and Kant. (It means essentially the transformation of philosophical discourse that blocks the direct presence of an integral human person in this discourse and allows for his/her presence only in some dismembered or disguised forms.) From the chosen angle, Hegel's System represents the extreme point of the anti-anthropological trend, and the philosophy of Hegel's opponent, Kierkegaard, the starting point of the opposite trend, the return to man. The beginning and the end (up to now) of this trend are given the main place: the anthropologies of Kierkegaard and the later Foucault are reconstructed in detail. In the interval between them we only consider Nietzsche and Heidegger (since no anthropological retrospection is thinkable without them) and Scheler (just briefly, as an interesting example showing how the chances opened for anthropological thought by phenomenology can be missed completely).

This scheme with its key conception of the "anti-anthropologism" emerged before I could read Foucault's *Hermeneutics of the Subject*, and I was glad to discover in this fascinating work close parallels to this conception. Foucault develops the idea about "two competing lines" in the history of European thought, to which he gives various names: the lines of cultivating, respectively, the care of the self and knowledge of oneself or the lines of "spirituality" and "philosophy," or else the lines of "cathartics" and "politics." He describes the gradual superseding of the "spirituality" line by the "philosophy" line in modern history (considering Descartes and Kant as the pivotal figures in this superseding) and then the inversion of this process, "the restoration and the second coming of the structures of spirituality," in the nineteenth and twentieth centuries. Clearly, in our scheme the "curve of the evolution" of European thought is depicted roughly in the same way.

What is even more important, both *Diogenes' Lantern* and Foucault's studies in the history of the subject are in no way restricted to purely historical goals. The theory of practices of the self, which is the base of Foucault's studies, is in fact oriented to the present-day situation, and the analysis of the past anthropological experience opens new prospects on how to solve contemporary problems of the human being and to formulate contemporary strategies of humane sciences. On the other hand, this book and all of *Diogenes' Lantern* are part of synergic anthropology, an anthropological approach that I have gradually developed over many years. Following the critical line in European philosophy of the twentieth century, this approach finds it necessary to reject the essentialist discourse of classical philosophizing based on fundamental metaphysical categories of essence, substance, and subject. Instead, it chooses to describe the human person in terms of human manifestations and practices based on the principle that the constitution of the human person is formed up in the experience of what I call the "anthropological unlocking," in which one's encounter with the Other takes place. The name of the approach refers to the ancient prototype of the unlocking principle, the notion of synergy, *synergeia*, in Byzantine theology. Synergic anthropology will be presented systematically in Part II of the book, and we shall make it clear that, like Foucault's theory, it tries to reflect afresh upon the basics of anthropology and to make them fit the reality of present-day humans with its above-mentioned tectonic shifts.

Moreover, both Foucault and synergic anthropology follow a similar strategy: reviewing a large field of anthropological experience, they single out a certain kind or formation of this experience and use it as the basis for developing a new nonclassical anthropology, the core of which is a new vision of human personality and subjectivity replacing the classical subject figure. However, the choice of the basic experiential formation is quite different: in Foucault, it is practices of the self and, in the first place, the practices of the Hellenistic époque of the first two centuries CE, chiefly, of the late stoics; in synergic anthropology it is spiritual practices and, in the first place, the mystical and ascetic practice of Eastern Christianity (Orthodoxy), hesychasm. Concepts and methods used in the analysis of these formations as well as the resulting theories and their recommendations for the present-day situation are no less different. In sum, the two anthropological approaches have a rich collection of both shared positions and sharp divergences.

Thus the genre of this book is comparative and dialogical. The first task of the book is the discussion of Foucault's theory, and so the first part is the detailed reconstruction of this theory in its phenomenal base, its ideas and structures.[1] The set of its basic concepts such as the self (*le soi* as different from *je, ego*), practice of the self, care of the self, conversion, *parrhesia*,[2] and so on, is composed and analyzed. Main classes of practices of the self such as spiritual exercises, practices of confession, the triad dietetics, economics, erotics, are discussed in both their conceptual and historical aspects. The semi-utopian part of the theory, the program of the "esthetics

---

1. The reconstruction is based on the corpus of texts that were available to me in 2008-2009 when the book was written.
2. *Free or bold speech* (Gr.), a verbal practice in Late Antiquity singled out by Foucault as especially important for the antique culture of the self.

of existence," presented sketchily by Foucault in the last years of his life, is reconstructed systematically and discussed critically. Discussion of methodological and epistemological principles of the theory makes it possible to position it in the modern context of the humanities.

But the core of Foucault's theory is his "hermeneutics of the subject," a new anthropological discipline presenting his vision of human personality and subjectivity. We specially concentrate on specific modes of subjectivity found by Foucault in his three principal formations or models of practices of the self: the Platonic model, the Hellenistic model, and the Christian model. A close view of the Foucaultian repertory of anthropological formations and structures of personality discovers soon that Foucault's theory lays the foundations of a new daring project of nonclassical and pluralistic anthropology. And here the book switches over to its second, comparative task: to the comparison of Foucault's project and its above-mentioned repertory with synergic anthropology and its ensemble of anthropological formations and paradigms of human constitution. I conceive such comparison as a dialogue with the late thinker, and indeed, it is sometimes a hot and polemical dialogue!

The basis for the dialogue is provided by the fact that, as already said above, the two projects have substantial common ground. First of all, on the grand scale they both are projects of roughly the same and rather rare type: they present a new vision of what is anthropology and what role it should have in the ensemble of the humanities. Thus their position in the context of modern humanistic knowledge is similar. After having declared the "death of man" in his early masterpiece, Foucault in his late conception of practices of the self performs (surprisingly without any rejection of his former views) an anthropological turn in humanistic knowledge: he puts the study of a certain class of anthropological practices into the center of this knowledge, and all his hermeneutics of the subject are aimed at giving an answer to the cardinal anthropological question of modernity: *Who comes after the Subject?*[3] As for synergic anthropology, it starts with a phenomenological description of a certain kind of anthropological experience (the experience of hesychasm and other spiritual practices), but then it extends its phenomenal base to all the area of extreme anthropological experience and finds out a complete set of paradigms of the constitution of

---

3. This formula has been coined by Jean-Luc Nancy as the title for an important collective work initiated by him: Eduardo Cadava, Peter Connor, and Jean-Luc Nancy, eds., *Who Comes after the Subject?* (New York and London: Routledge, 1991).

the human being as such. Here it leaves the ground of religious experience and develops gradually into a general anthropological conception, which unfolds phenomenological description of the whole field of anthropological reality as such. Its eventual relation to religious experience and religious discourse is somewhat similar to the case of existential psychotherapy by Ludwig Binswanger (whose work influenced Foucault's early period). Indeed, Binswanger characterized his approach as follows: "I am not going to develop Christian anthropology from the Christian point of view, but I work as far as possible in a purely anthropological framework, trying at the same time to show how the religious sphere is included into the essence of the anthropological."[4] In our case, however, the distancing from religious discourse is more substantial. Synergic anthropology relates the religious sphere (more precisely, spiritual practice) not to the "essence of the anthropological" (a concept it rejects), but to just one of the basic paradigms of the human constitution — also describing and studying other paradigms in which such relation is absent.

Basing itself on the set of paradigms of the human constitution, synergic anthropology launches the program of a far-reaching "anthropological expansion" that includes in its orbit social reality and advances to an anthropologizing reconceptualization and reinterpretation of all humanistic discourses. One easily agrees that the anthropological turn performed by Foucault's theory moves in the same direction; it can also be considered as an expansion of anthropological discourse (represented as the discourse of anthropological practices) to the fields of history, social and political philosophy, psychology, etc. — in principle, to all the fields of modern humanistic knowledge. Thus both projects advance to a new image of anthropology: anthropology conceived as a meta-discourse with respect to the whole set of humanistic disciplines and discourses or a new *epistēmē* for humanistic knowledge, or else a *science of humane sciences*. Such meta-discourse differs from philosophical, cultural, social, and other known kinds of anthropology, treating them as particular cases or "partial" anthropologies that restrict their subject field to one or another part of integral anthropological reality. To my mind, the creation of such an anthropology is the principal task of modern philosophical and anthropological thought.

Self-evidently, synergic anthropology rooted in hesychast experience and Eastern Christian discourse, and Foucault's theory with its completely

---

4. Ludwig Binswanger, "The Letter to Semyon Frank of 11 November 1936," in *Semen Lyudvigovich Frank*, ed. V. N. Porus (Moscow: ROSSPEN, 2012), p. 482.

secular roots and basis, have also many divergences including some fundamental ones. There is no need to concretize them here, as they are described in detail in Part II of the book. Synergic anthropology is not a theological or confessional discourse, but it considers religious experience or at least the experience of spiritual practice as a sound and full-fledged kind of anthropological experience, in which man unlocks himself ontologically and not just ontically, actualizing a certain important paradigm of his constitution. In no way is this the case with Foucault's theory. Foucault the thinker inherited the long French tradition of hostility to religion, which was enhanced in him by certain personal factors. But nevertheless the significance of this ideological motive must not be exaggerated. Contrary to it, Foucault's theory and synergic anthropology share a number of common basic features. Indeed, as this book shows, both Foucault's theory and synergic anthropology are based on nonclassical and nonessentialist principles, they reject a Cartesian conception of the subject and Cartesian-Kantian anthropology and epistemology, and they characterize anthropological reality not by abstract categories, but by anthropological practices. But in the case of synergic anthropology, all these features can be traced back to the Eastern Christian discourse and recognized as its characteristic distinctions. As such, they can be found in varying forms in works of modern Eastern Christian philosophers and theologians, e.g., Metropolitan John Zizioulas, Christos Yannaras, and George Mantzarides. And it means that Foucault's theory has considerable common ground not only with synergic anthropology, but also with the Eastern Christian discourse as a certain spiritual and intellectual formation.

The last conclusion may seem paradoxical if not absurd. The common ground, the closeness between the positions of a Parisian leftist, a gay-movement partisan, and an ancient religious tradition based on monastic spirituality and an intricate school of prayer? What? Are you serious? Yes I am. In many respects the relation of Foucault's theory and Eastern Christian discourse gives us a good example of *coincidentia oppositorum*, proving that the meeting of sharply different worlds can produce not only conflicts and clashes, but also new fruitful connections. To my mind, such meetings are symptomatic nowadays and they will become more and more typical in postsecular configurations that are now forming up in global political, social, and cultural reality. In postsecular dialogue many old ideological labels and barriers are removed, including, in the first place, the barriers between secular and religious consciousness, thought, spheres of social and cultural life. It is a positive trend because such dialogue helps one to reduce growing

PREFACE TO THE AMERICAN EDITION

threats and risks in the present-day global development. It increases the chances for mutual understanding and survival of all of us in the global village — and I hope that American readers will find in this book, besides its academic message, a certain contribution to this dialogue.

SERGEY S. HORUJY

Introduction

I have long noticed in myself a lingering effect from reading Foucault of the late period: for some reason there would pop into my mind the pages of *Safe Conduct*[1] where Pasternak describes a phenomenon he calls "the last year of the poet." Looking at the finales of a number of poets' biographies that are of particular significance for him, he indicates that these finales are characterized by an unprecedented intensity of activity and saturatedness with meaning, by a power of creativity that subordinates everything to itself and which is full of a mighty *drive* that forcefully carries one along in its striving toward some higher culmination — toward an unknown but predetermined synthesis, toward an outwardly invisible result . . .

I am by no means a Foucault fanatic or a specialist in his work, and therefore I did not devote too much thought to whether there was any validity in this impression of mine or to what caused it. But then there appeared the Russian edition of *Hermeneutics of the Subject*, the lecture course of 1982, and it became clear to me from the article of its editor, Frédéric Gros, that my vague impression was indeed valid. The final period of this philosopher stood out here in a bright light, and it became perfectly clear that there was indeed a "last year of the poet" in Foucault's life. This is what Gros writes: "The last years of Foucault's life, from 1980 to 1984, were a period . . . of ever-increasing intensity . . . a period of startling acceleration of intellectual work, a burst of creative activity. Nowhere else does one feel so strongly that which Deleuze calls rapidity of thought."[2]

---

1. An autobiographical work (1931) by Boris Pasternak. — Trans.
2. Frédéric Gros, "O kurse 1982 goda" (On the lecture course of 1982) in Michel Fou-

INTRODUCTION

I have mentioned this special period of Foucault's creative activity not in order to make biographical or existential observations. What is important for me is its essence. This period is also extraordinary in terms of the ideas and principles promulgated in it: we find in it a decisive change of landmarks — a change in the themes and problems considered, in the fundamental concepts and points of view. The scale of the changes was so great that Gros considers them to be a veritable "conceptual revolution." The essence of the revolution consists in the fact that Foucault's philosophy becomes a *theory of practices of the self*: its main theme is the immanent constitution (and, in diachrony, the "genealogy") of the subject, and its main working concept is "practices (or technologies) of the self," that is, practices of the self-transformation of the subject instead of practices of power and practices of knowledge, which had formerly been in the forefront. Foucault acknowledges that these practices are constitutive for man and irreducible to any other type of his practices. This means that the "conceptual revolution" also contains an *anthropological turn*. Foucault takes the positions of the irreducibility, autonomy — and possibly even of the primacy? — of the anthropological level of reality; and his theory of practices of the self can rightfully be considered an essay in nonclassical anthropology.

Meanwhile I had long been investigating in Russia a certain specific class or ensemble of practices of the self — the practices developed in the Orthodox mystico-ascetic tradition of hesychasm known also as sacred silence or noetic practice (*praxis noēta*, in Greek). I had reconstructed the original practical anthropology created in the school of hesychast ascesis, and after analyzing the principles of this anthropology and making generalizations, I had arrived at a general "paradigm of spiritual practice" describing the anthropological foundations of the mystico-ascetic practices, Western and Eastern, created in the world religions. These, too, are practices of the self, but of a special and distinctive type, for these are practices of *the constitution of man in his ontological unlocking*. In turn, the analysis of spiritual practices led to an even broader class of *extreme practices of the self*, in which other types of the unlocking of man are realized and which, in their totality, form what I have called the "Anthropological Border" or

---

cault, *Germenevtika sub'ekta* (Hermeneutics of the Subject) (St. Petersburg: Nauka, 2007), p. 558R, p. 563R [Russian translation]; p. 495F, p. 499F. All quotations from this edition have been checked against the French original. Page numbers from the Russian translation are followed by the letter "R"; page numbers from the French edition used (*L'herméneutique du sujet* [Paris: Seuil/Gallimard, 2001]) are followed by the letter "F." — Trans.

Introduction

the "energetic border of Man."³ These extreme practices are used to formulate a new method for describing anthropological reality called *synergic anthropology* (from synergy, the notion of Byzantine theology meaning the harmonious, coherent action and cooperation of two energies belonging to different sources). Like the theory of practices of the self, synergic anthropology represents a particular approach to the hermeneutics of the subject, and, as is already clear to some extent, these two approaches, placing the constitutive practices of the self-transformation of man at the center of the investigation, exhibit numerous mutual correspondences and profound internal interconnections.

To my mind, Foucault's theory of practices of the self is the most significant contemporary essay in anthropological thought; perhaps it is also the first substantial outline of a new anthropology. In what follows, the analysis of this theory will be conducted in the framework of a systematic comparison with the positions of synergic anthropology. The parallel examination of these two new approaches will naturally bring us to reflections on the present-day anthropological situation and on the final Pushkinian question, "So where shall we sail?"

---

3. See, e.g., Sergey S. Horujy, *Ocherki sinergijnoj antropologii* (Essays in Synergic Anthropology) (Moscow: Institut filosofii, teologii i istorii Sv. Fomy, 2005) [in Russian]. In English, see the article: "Man's Three Far-away Kingdoms: Ascetic Experience as a Ground for a New Anthropology," *Philotheos. International Journal for Philosophy and Theology* 3 (2003): 53–77.

I

Foucault's Last Project,
or: Hermeneutics by No Means of the Subject

"The technology of the self" is an enormous and very complex domain whose history it is necessary to create.[1]

The first step in this investigation is to sketch the simple external framework of the phenomenon that interests us, that is, of the theory of practices of the self (or rather of the program or project of these practices, since nothing like a systematic theory of them was formulated). Foucault preferred (which is natural) to view the development of his thought in terms of continuity, not discontinuity; as Gros says, "[E]very thought which he presents to us as new he also finds in an undeveloped form in his earlier works." All the same, the boundaries of the new project are very clearly defined.

All of Foucault's major works that were formative for his philosophy, from *The History of Madness* (1961) and *Words and Things*[2] (1966) to the first volume of *The History of Sexuality* (1976), were focused on practices of power and discursive practices, were characterized by the primacy of a social-institutional approach, and tended "to conceive the subject as a certain objective derivative of systems of knowledge and power," as "a passive

---

1. Michel Foucault, "A propos de la généalogie de l'éthique: un aperçu du travail en cours," in *Dits et écrits* II, 1976-1988, no. 344 (Paris: Gallimard, 2001), p. 1447. Further citations from this edition are given as *DE II*, with the text numbers and page numbers indicated. All quotations from *DE II* have been translated from the French text by both the author and the translator.

2. Published in English as *The Order of Things* (1970). — Trans.

product of technologies of domination" (Gros). This also defined how the cultural community viewed Foucault: the large conference on Foucault's work held in Los Angeles formulated his main themes as "Knowledge. Power. History." But at the time of this conference (October 1981) such a formulation was far from fully corresponding to reality, as Foucault himself announced in his interview with *Time Magazine:* "I am interested not so much in power as in the history of subjectivity." At the beginning of 1980 Foucault reads a lecture course at the Collège de France on *The Government of the Living*, the main theme of which is practices of the self in early Christian ascesis. This lecture course (still unpublished, alas) is the first landmark of his new period. This is what Foucault says about it in the 1982 lecture course: "It was only at that time that I was beginning to occupy myself with these things" (i.e., with practices of the self); and Gros calls this "the first deviation from the defined path," i.e., from the original program of *The History of Sexuality*, a program that is still based on the old principles. The whole of 1980 represents a landmark or boundary: this year is "decisive for Foucault's intellectual path"; "this is a time of the problematization of the technologies of the self as reducible . . . neither to technologies of the production of things, nor to technologies of power over people, nor to symbolic technologies."[3] And this is precisely the governing principle of his new period, indicating the path and method of "conceiving the subject in a new way."

Thus, the chronological frame for my investigation is 1980-84. This is quite a short period if one takes into account the fact that during all this time Foucault also concerned himself with politics, taking part in protests against various political persecutions (of the Solidarity movement in Poland, of the Senegalese in France, of Vietnamese refugees), and that the last two years before his death of AIDS on 25 June 1984 were years of great physical suffering. But before us we have "the last year of the poet," and his creative output during that period was astonishing. I will give only the briefest of lists. Foucault manages to submit for publication volumes 2 and 3 of his *History of Sexuality*, which are now written on a new conceptual basis; volume 4, devoted to the early Christian epoch, went through all but final editing and remains unpublished. He did not skip a single year in his lecture courses, which comprise:

---

3. Frédéric Gros, "O kurse 1982 goda" (On the lecture course of 1982) in Michel Foucault, *Germenevtika sub'ekta* (Hermeneutics of the Subject) (St. Petersburg: Nauka, 2007), p. 570R; p. 505F.

1981. *Subjectivity and Truth.*[4] (Its theme is pleasure in Greco-Roman culture of the first to second centuries. Material from the course was incorporated in volume 3 of *The History of Sexuality*.)

1982. *The Hermeneutics of the Subject.*[5]

1983. *The Government of Self and Others.*[6] (Its theme is the practice of free speech, *parrhēsia*, in ancient Greece.)

1984. *The Courage of the Truth.*[7] (Its theme is the practice of free speech in Hellenistic and early Christian culture.)

His growing fame leads to a multitude of workshops devoted to his theories; he participates in a number of them, presents lectures, and writes articles. Here is a list of some of the most significant of these events:

1980, October-November. "Christianity and Confession," lectures at Berkeley and Dartmouth College.

1981. "Sexuality and Solitude," lecture in New York; there is an important draft of it (1980).

1982, June. "To Speak the Truth about Oneself," paper presented at the International Summer School for Structural and Semiotic Studies in Toronto.

1982, October-November. "Technology of the Self," workshop at the University of Vermont.

1983, February. "Self Writing," an article.

1983, April-May. Series of lectures, discussions, and interviews on the theme of practices of the self at Berkeley.

1983, October-November. Workshop on practices of free speech at Berkeley.

1984, 29 May. "The Return of Morality," an interview.

What are the most important works in this abundant material and what will serve as the basis of the present analysis? Most essential for our purposes are texts that formulate the conceptual foundations of the theory of practices of the self as well as texts on practices of the self cultivated in Christian culture, these latter texts being used for comparing Foucault's ideas with my own conceptualization of spiritual practices.

4. Published in English translation in 2007. — Trans.
5. Published in English translation in 2005. — Trans.
6. Published in English translation in 2010. — Trans.
7. Published in English translation in 2011. — Trans.

As far as the first type of texts is concerned, one cannot find in Foucault — looking at the character of the entire oeuvre of this thinker as a *historian of thought* — any systematic development of the "conceptual foundations" that interest me here. He created his own special discourse, which indissolubly fuses history and philosophy, as well as ethics and politics; and all the conceptual elaborations are integrated into historical studies. He insisted that this methodology, which refused "to accept from historians in a ready-made form that upon which one must reflect," is "the only way . . . to keep from being enslaved by the hidden postulates of history"; and he called this methodology "reflection in history."[8] However, the conception of practices of the self is not only a historical conception, but also a rich general anthropological conception that requires its own conceptual apparatus and generates a whole spectrum of philosophical and methodological problems. By no means is Foucault unaware of this general problematic, but in the framework of "reflection in history" his thought always also poses for itself tasks of a historical order, so that, in sum, every one of his books and every one of his lecture courses has its own peculiar balance, its own formula of the union of synchrony and diachrony. In this respect the lecture course of 1982 represents a salient point. Although here too we have a substantial historical task, a reconstruction of the practices of the self of the Hellenistic period, nevertheless the theoretical task — the creation of the foundation and apparatus of the conception of practices of the self — is just as important and receives the same degree of attention. It would not be wrong to say that this course presents not only an examination of practices of the self in late antiquity but also a clear framework for the general conception of practices of the self, even though this framework is still far from being a "full-fledged theory." *It is precisely* The Hermeneutics of the Subject *that will be the basic text for my analysis.* In addition, in the volumes of *The History of Sexuality* that have been published, besides sporadic comments on the properties of practices of the self, there are sections focusing on these properties: the introduction to volume 2, where a reorientation of the whole project of the *History* is announced; and in volume 3, the second chapter, "The Culture of the Self" and (to a lesser extent) the third chapter, "The Self and Others." But all the elements of the conceptualization of practices of the

---

8. Foucault, "A propos des faiseurs d'histoire," *DE II*, no. 328, p. 1232. *Mutatis mutandis*, the principle of "reflection in history" can be methodologically compared with the principle of "philosophizing in religion after immersing oneself in its element" put forward by Florensky at the beginning of *The Pillar and Ground of the Truth*. Both thinkers summon one to incorporate into philosophy itself the experience that grounds it.

self revealed in the *History* can, as a rule, be found in *The Hermeneutics of the Subject* in a more detailed form.

As for the material on practices of the self in Christian ascesis, and more broadly in Christian doctrine and culture, it is not so much meager as largely inaccessible, alas, for the most part. Foucault devotes to these practices volume 4 of *The History of Sexuality* ("Confessions of the Flesh"), the 1980 lecture course, *On the Government of the Living,* and (in part) the 1984 course, *The Courage of the Truth*. All these sources are as yet unpublished; accessible to us are only the "Brief Summary" of the 1980 course and a single fragment of "Confessions of the Flesh," which the author published as an article in the collection *Western Sexualities* (1982). Apart from this, on this theme there also exist the two concluding sections of the workshop lecture "Technologies of the Self" (1982, Burlington, VT), as well as two lectures, from 19 and 26 February, on practices of repentance and confession, in the 1974-75 course titled *Abnormal,* given in the period before the "conceptual revolution." Despite this limited material, it is nevertheless possible to form a fairly clear idea of Foucault's positions vis-à-vis this theme, at least with regard to the main points. A scholar has the right to have predilections, and Michel Foucault, from the whole history of the practices of the self cultivated in the West, chose and fell in love with a single period — the period of Hellenistic culture of the first to second centuries. He called this period the "Golden Age of practices of the self" and gave it his foremost attention; he fully devoted to it, in particular, the 1981 and 1982 courses and volume 3 of *The History of Sexuality*. But the Christian world is the direct successor, and in part the contemporary, of this "Golden Age"; and in his analysis of the practices of the self, as well as of the concepts, orientations, and principles of behavior in the culture of late antiquity, the fundamental methodological approach used by Foucault is their comparison and contrast with Christian practices (concepts, orientations, principles, etc.). He uses this approach constantly, with application to all his themes, and as a result there also emerge before us his assessments and interpretations of all of the most important phenomena of the Christian culture of the self. *The Hermeneutics of the Subject,* supplemented by some smaller sources, will give us sufficient material.

I.1. The Language of Foucault's New Conception of the Subject

It is important to first get a grasp on Foucault's language of description, on the basic terminology of his new conception of the subject. What exactly do

PRACTICES OF THE SELF AND SPIRITUAL PRACTICES

"the care of the self," "the practices of the self," and "the culture of the self" mean? — these terms which, after Foucault, rapidly and without particular reflection gained wide currency in the humanities.

Let me begin with the central working concept, which is used to designate the conception as a whole. The *practice of the self* receives in Foucault a whole series of definitions. These are two of them: Practices of the self are "certain procedures which indisputably exist in every civilization and which are proposed or prescribed to individuals for establishing their identity and for maintaining or transforming it. . . . [T]hey are possible thanks to relations of self-control or self-knowledge."[9] They are also "intentional and reflected practices by means of which people not only establish for themselves rules of behavior, but also seek to transform themselves, to change themselves in their unique being, and to make their lives into works that have a certain esthetic value."[10] The first definition refers more precisely to "technologies of the self," a term that is nearly, but not completely, synonymous with "practices of the self." Simplifying to some extent, one can say that practices of the self are technologies of the self that possess an important additional property: they are oriented — perhaps not directly but in the final analysis — toward giving an individual access to truth. This aspect is expressed, for example, by the following one of Foucault's formulas (it serves for him as a definition of "spirituality" but can also be applied to practices of the self): practices "by means of which the subject produces in himself changes necessary for gaining access to truth, they can be practices of purification, ascesis, renunciation, conversion, or change of one's mode of life" (27R; 17F).[11]

Clearly, these three formulas contain a great deal that is indefinite and not fully expressed. They have only one common denominator: there is no doubt that "the practice of self" is a practice of the self-transformation of the subject. But each of the formulas used by Foucault characterizes these

---

9. Foucault, "Subjectivité et vérité," *DE II*, no. 304, p. 1032.

10. Foucault, "Usage des plaisirs et techniques de soi," *DE II*, no. 338, p. 1364.

11. Here and elsewhere the numbers in parenthesis after quotations indicate page references to *The Hermeneutics of the Subject* in the Russian translation: *Germenevtika sub'ekta* (St. Petersburg: Nauka, 2007). Let it be noted that sometimes the quotations are not from the 1982 course but from unpublished texts of Foucault copiously quoted by Frédéric Gros in the supplementary materials to this edition. [All quotations from this edition have been checked against the French original. Page numbers from the Russian translation are followed by the letter "R"; page numbers from the French edition used [*L'herméneutique du sujet* (Paris: Gallimard, 2001) are followed by the letter "F." — Trans.]

transformations in a very different way, and it is not at all clear that these characteristics fully agree with one another. Furthermore, we find in them some highly loaded philosophical terms that force us to commit ourselves: "subject," "truth," "identity," and the nontraditional, albeit not new, term "self" *(soi)*. How, in their turn, should we understand them? Carefully analyzing the usage of concepts in the historical contexts that are studied, Foucault's late texts avoid immersing themselves in the modern problematic of the foundations of philosophical discourse,[12] and one finds in them only some fleeting guiding remarks, but by no means a finished treatment of the above-mentioned concepts as well as other concepts of this discourse. What is "access to truth," an expression that presupposes — contrary to the positions of classical epistemology — that the possibility of attaining truth requires some sort of special anthropological premises in addition to the proper process or act of knowing? There are many other questions that can be posed. Clearly, Foucault does not give a complete philosophical conceptualization of the phenomenon — yet one can already see from his formulas why practices of the self can become the central and key concept of a new method for describing man.

From the first and the last formula we conclude that practices of the self, in distinction to simple technologies of the self, have a teleological element: these are the *directed* transformations which are determined by a certain pre-specified general-anthropological goal. In the one case this goal is designated as the constituting (acquisition, protection, change) of man's identity, and in the other case as the assurance of "access to truth." This formulation is obscure, but Foucault clarifies it: he has in mind "the transformation of the subject with the goal of making him conform with truth" (28R; 18F) — in other words, transformation into a certain state that "conforms with truth." This is a very special state — one that "imparts completion to the subject ... and allows him to be realized" (29R; 18F). In this state he clearly emerges as the "truthing subject" or "true subject," in other words, as his "true self." The becoming "realized" of a subject, his "acquiring of completion," is *eo ipso* the constituting of the subject; and we conclude that in this case the goal and essence of the practice of the self are indicated also in the constituting of the subject, of his personhood and identity, and here the constitution is additionally characterized by its relation to truth: the constituting of the subject is his

---

12. Nevertheless, we can find scattered remarks which indicate that Foucault took a rather definite position in this problematic, and indeed his whole "new course" promised to lead inevitably to a weighty new word in it.

becoming his "true self" (or, what is the same thing, his "attaining access to truth," his becoming "conformable with truth").

The aforesaid is sufficient to allow us to arrive at a preliminary conclusion. Practices of the self, according to Foucault, are anthropological practices of a kind in which the constituting of the subject is realized; they are *constitutive,* and as a result they are capable of occupying a central place in the anthropological discourse; on the basis of them one can develop an integral description of anthropological reality. A large problem follows from this — one of the fundamental problems connected with unfolding the conception of practices of the self: namely the problem of bringing to light the new paradigm of the constitution of man imbedded in this conception. It is immediately obvious that this paradigm is *nonclassical,* i.e., that it differs from the analogous paradigm of classical European anthropology, which measures man by the concept of his essence or nature and presupposes that he is constituted in the process of the actualization of his essence. Foucault's conception is not based on the idea of essence or nature. One can also expect that, by virtue of the great diversity of the possible modes of man's relation to himself, of the strategies and goals of self-transformation, such a paradigm is a priori not unique. In synergic anthropology and the spiritual practices it describes, the nonclassical paradigm of constitution is realized in the form of the *paradigm of unlocking.* It presupposes that man is constituted in the experience of unlocking himself — that is, in an experience in which man, reaching the boundaries of the horizon of his consciousness and existence, becomes open to that which is beyond this horizon and is formed by interacting with it (for more detail about this, see Section II). This kind of paradigm seems very natural as far as practices of the self and "becoming one's true self" are concerned, and yet Foucault sees in the Hellenistic practices of the self (the main class of practices studied by him) something different, something that is more like a locking than an unlocking of man. This is one of the specific points of his conception that must be clarified, and I will discuss it later on.

One more important aspect of the constitution of the subject as well as the relations of the subject to truth, is the fact that Foucault's treatment of these topics immediately reminds one of Kierkegaard.[13] The idea of "access

---

13. In my reconstruction of the evolution of European anthropology — *Diogenes' Lantern* (Moscow: Institut filosofii, teologii i istorii Sv. Fomy, 2010) [in Russian] — Kierkegaard appears as the author of the first "nonclassical" anthropological model not using the concept of the essence of man.

to truth" is wholly based on the thesis: "Truth is not given to the subject by a simple act of knowing . . . it is necessary that the subject change" (28R; 17-18F). This thesis is a variation on one of the main themes of Kierkegaard, who wrote, for example, that "[I]n the spiritual world every change of place signifies a change of the traveller himself" (in the *Unscientific Postscript*). In resolving the problem of constitution there is even a more direct resemblance with Kierkegaard. As early as in the first of his major works, *Either/Or*, Kierkegaard developed a conception according to which a human being is constituted by "choosing himself" and becoming his true self. Foucault reproduces closely this model of constitution, and it even becomes the very core of his whole conception of practices of the self. However, he does not mention Kierkegaard, even when he enunciates theses that literally sound like Kierkegaard's (cf. for example: "The goal of the practice of the self is I myself. Only a few people are capable of being themselves" [147R; 122F]). This is only the first case of a close and deep connection of Foucault's ideas with Kierkegaard; later on we will see quite a few such connections, and yet Foucault never mentions the Danish philosopher.

It is clear that the practices of the self become the *topos* of Foucault's new conception of the human subject, the nest in which is collected the extensive ensemble of his concepts and ideational threads. To start with, it is necessary to examine the "internal" notions of the topos that reveal the content of practices of the self as an anthropological paradigm. Logically, the *telos* of the practice must be the most important of these notions. When the practice of the self is teleological, when it is directed at achieving a certain final state of a human being, which state is called the *telos*, this *telos* plays a special defining role in the whole process of the practice: every phase of the practice, every state of the human being attained in the course of the practice, is characterized and verified by its relation to the *telos*, and by the "distance" that separates it from the *telos*. Such is the role of the *telos* in spiritual practices. But we virtually do not encounter this notion in Foucault's conception, and he himself explains the reason for this. The main object of his investigation is the practices of the Hellenistic epoch; he introduces his general concepts based on the material of these practices and with application to them. By no means does he deny the cardinal significance of the *telos* in the structure of practices of the self; however, in the Hellenistic practices, as he discovers, the *telos* remained insufficiently reflected upon and not fully defined. In the capacity of the *telos* there appears, as we have seen, a certain "true self" *(le soi)*, "the self which has become conformable with or open to truth," "the self itself" *(auto to auto)*, and so on. Needless to say, these formulas are obscure and can only give us hints; and, according to Foucault, in Hellenistic culture no *bona fide* concept has ever been made out of them: "... Hellenistic and Roman thought was not able to clarify and decide whether 'the self' *(le soi)* represents something to which one returns because it was given in advance, or whether it is a goal which one must set before oneself and which one can eventually attain.... I believe that here, in this practice of the self, we have a case of fundamental uncertainty, of fundamental doubt" (240R; 205F). But avoiding, like the Hellenistic practices, the concept of *telos,* Foucault's discourse of practices of the self also avoids and leaves indeterminate the whole spectrum of problems that have any relation to this concept; thus, Foucault gives us only fragmentary glimpses of the ontological content of the practices of the self as well as of the whole general ontological problematic. We find here an important distinction from the discourse of spiritual practices, a point I will return to in Section II.

The next internal characteristic of practices of the self is what I call *holisticity.* The theme of practices of the self is in part developed by Foucault in a polemical key: he affirms the importance and autonomy of such practices in relation to those that had previously served as his cornerstone — in relation to practices of power and practices of knowledge. The clearest distinction from practices of knowledge consists in the fact that all the levels of man's being are incorporated into the practice of the self. The degree and concrete character of the co-participation of the bodily and emotional spheres in the practice of the self change drastically in different epochs, but the very fact of this co-participation is an invariable feature of the practice. In virtue of this *predicate of holisticity* the practice of the self represents a "complex interweaving of the psychical and the bodily." This interweaving is realized in totally different ways in the three "major forms" of the practices of the self in the West (Platonic, Hellenistic, and Christian), and therefore we will speak about this holisticity more concretely below, when we discuss these "major forms."

However, perhaps the only universal property of this predicate for Foucault lies in the fact that he describes the heterogeneous structure of the practice of the self generated by the holisticity of the practice as a combination of two spheres designated by the ancient terms *Mathesis* and *Askesis* and containing, respectively, cognitive activities and activities of action, or exercise. It is important to note one aspect of Foucault's treatment of *Askesis:* he asserts that this sphere is outside the influence of the laws of society, is governed not by these laws but directly by the goal of "linking the subject with truth" — in other words, by the *telos* of anthropological practice. "Ascesis," and with it the practice of the self, is thus taken out of the sphere of social phenomena; its nature is not social but purely anthropological. The task of the practice of the self (i.e., "the establishment of the subject as the final goal for himself") and consequently the very constitution of the subject, is posed therefore as a properly anthropological task; the subject is no longer constituted from the social sphere — he is no longer a "passive product of technologies of power." Thus, this detail represents a clear sign of Foucault's new positions, a sign that the "anthropological turn" has been accomplished. In speaking about his beloved Hellenistic culture, he affirms even more clearly and decisively the existence of an autonomous authentically anthropological sphere of the constitution of the subject: "Neither the structure of the polis, nor the laws, nor the demands of religion could ever dictate to the Greek or to the Roman, but primarily to the Greek, how he should conduct himself in various life-situations. And, most importantly, they were not capable of explaining to him how he should build his life." Therefore the practice of the self "occupied in Greek culture a place that remained unaffected by the regulatory activity either of the polis and the law or of religion" (485R; 429F). The practice of the self was a domain of man's free self-actualization.

The necessary presence of the transindividual dimension, *the participation of the Other*, is the next internal characteristic of the practice of the self that we will examine. This is an absolutely universal feature of the anthropological paradigm; Foucault asserts its necessity when he discusses (a rare occurrence) precisely the practice of the self in general or as such: "The other is indispensable in the practice of the self if the form defining this practice is to really attain its goal and reach its object which is the self *(le soi)*, and fill itself with the latter. For the practice of the self to arrive at this self which is its goal, the other is indispensable" (148R; 123F). But this thesis (which, strictly speaking, is a negative one, since it declares only that the subject by itself is insufficient for the success of the practice) essentially exhausts the universality of the "discourse of the Other." One can progress a step further by characterizing the role of the Other in sufficiently universal terms: The Other "is a specialist in restructuring the individual and in forming the individual into a subject. He is the mediator in the relation of the individual to his constitution as subject" (150R; 125F). However, why is such a specialist needed? Why can the subject not be "the operator of his own transformation?" (150R; 126F)? Why can he not actualize or embody the form of the practice of the self? Finally one must show concretely *what is lacking in the isolated subject* of the practice, in the subject who is confined within the limits of his individuality, of his "singularity"! And here the answer is no longer universal; it sends us back to the concreteness of cultural epochs and changes as a function of this concreteness (so that the "discourse of the Other" thereby becomes an empirical discourse, which does not attain philosophical conceptualization).

In the original Platonic form, where the practice of the self was not yet fully constituted, the only thing that the subject, the youth, lacked was knowledge and a personal example, a model; as Foucault says, here the activity of the Other, the Mentor (whom he calls a "psychagogue"), has not yet separated itself from pedagogy. In the Hellenistic form psychagogy is necessary in order to take the subject out of the ordinary psychic structure of routine life, which is judged to be pathological, so that here the Other, the Teacher, is likened to a physician (the medical parallel is one of the main themes of Foucault's reconstruction of the Hellenistic practice of the self). "A human being always excessively loves himself" and he is therefore incapable of "being his own physician . . . when it comes to healing his passions and errors. . . . Nothing will be achieved without another's help" (429, 431R; 379, 381F). This resembles to some extent the spiritual directorship that evolved in Christianity; however, the differences between this conception and the Christian care of souls, and in particular its specific form in ascetic practice, are profound and fundamental. (I will touch upon these differences later on.) There are even greater differences when it comes to the concrete functions of the Other in different formations of the practice of the self.

Finally, the most important structural characteristic of the practice of the self is the paradigm of *conversion to the self*,[14] which discloses the general type of the practice as a certain "process of movement" performed by the individual. This is the central paradigm of Foucault's entire conception. As is usually thought, this conception had its origin in Foucault's encounter with the conversion paradigm: his attention was drawn to the latter by its elaboration in Pierre Hadot's works on "spiritual exercises" in the culture of late antiquity.[15] Later on, however, he developed his own detailed treatment

---

14. *Conversion à soi.* — Trans.

15. "Foucault told me once that he was influenced by the first of my articles which was devoted to the concept of 'conversion.'" Hadot said this in response to a question at an interview, in which the interviewer asserted more generally that "at the end of his life Michel Foucault became greatly interested in technologies of the self and in the practice of the self under the influence of your idea of spiritual exercise." Pierre Hadot, *Dukhovnye uprazhneniya i antichnaya filosofiya* (Spiritual Exercises and Ancient Philosophy) (Moscow-St. Petersburg: Stepnoy Veter, 2005), p. 367. In French: *Exercices spirituels et philosophie antique* (Paris: Nouvelle édition, 2002).

of conversion, and personal relations were established between the two philosophers that both of them found valuable. At the very beginning of "The Use of Pleasure," Foucault says that he "found the works of Peter Brown and Pierre Hadot very useful; conversations with them and their views helped me repeatedly." As for Hadot, in two articles written after Foucault's death he speaks of his "great fortune" that he was able to carry on conversations with Foucault and of the "interrupted dialogue" with him which was "nourished" by their disagreements. (Indeed, these two articles contain a great deal of criticism not only of Foucault's conception of conversion but also of his entire theory, and we will have occasion to refer to these articles more than once.) There is no doubt that although the lion's share of Foucault's treatment is reserved for late antiquity, the most universal meaning is attached to the paradigm of conversion. According to Foucault, conversion has a multi-aspect character: it arises in different forms and in the most diverse epochs, including our contemporary one; one finds it in practices belonging to various spheres of human self-realization, and the concept expressing it can belong to various discourses. "The concept of conversion... is one of the most important technologies of the self known in the West.... But it would be a great error to limit the importance of the concept of conversion to the sphere of religion alone... it is also an important philosophical concept and one which has played a decisive role in philosophy.... The concept of conversion also possesses a capital importance in the moral sphere. Finally one should not forget that since the nineteenth century this concept has inserted itself in a spectacular and even dramatic fashion into thought, into practice, into experience, and into political life" (234R; 199-200F). A vivid example of this universality of the paradigm is its presence, asserted by Foucault, in revolutionary practice, at various stages of the revolutionary consciousness.

Wholly devoting the first half of his lecture of 10 February 1982 to the idea of conversion, Foucault introduces this idea as a kind of deep substratum of the entire ancient culture of the self. After describing the psychological background and the cultural and social context of the practice of the self in its mature form in late antiquity, Foucault concludes: "One and the same impressive image — the image of the conversion to the self — can be glimpsed behind all these representations" (233R; 198F). It could not have been otherwise: The *telos* of the practice of the self is "I myself," the "self"; and this means that in order to attain it the individual must first achieve directedness or convertedness at himself. "We must turn away from everything... that is not ourselves... we must turn to the self, convert to the self

... our entire being must be turned and converted to ourselves" (232-33R; 198F). But it is also clear that this intuition of conversion to the self is far from containing a finished concept, and indeed such a concept cannot arise in the frame of a general anthropological discourse. Both in its nature and essence and in its actualization "conversion to the self" is, once again, defined by the *telos* of practice, by that "true self" to identification with which the conversion must lead. And, fundamentally, this *telos* is not universal! Different cultures and even different epochs of the same culture can differ cardinally in the constructions they create of the "self," of the "true self," etc. For the three "great formations" of the practice of the self considered by Foucault's theory (Platonic, Hellenistic-Roman, and Christian) these constructions also differ (and in addition, in the Hellenistic model, which is the main focus of Foucault's attention, the construction essentially remains somewhat vague). However, Foucault does make one bit of significant progress on a general level: he notices (following Hadot to some extent, however) that a bifurcation or alternative is imbedded in the paradigm of conversion. Depending on the nature of the "distance" or difference between the initial empirical "self" and the *telos* of the practice of the self which "conforms with truth," the essence of the conversion to the self can be of two kinds: first it can be a *return to the self* (if the "true self" is presupposed as having once existed but having been lost or forgotten or as being hidden in the depths of the empirical "self," and so on), or, secondly, it can be a drastic and fundamental change, a *transformation of the self* (if the "true self" has a different nature and is separated by a rupture from the empirical "self"). In the first case Foucault calls the change or event occurring with the individual in the conversion and practice of the self "self-subjectivation"; in the second case he calls it "trans-subjectivation." These concepts seem fruitful and useful for modern investigators of subjectivity, but unfortunately, after introducing them, Foucault does not develop them further. We will return to this bifurcation of conversion when we consider concrete formations of the practices of the self: as is clear beforehand, conversion-return corresponds to Platonism and Neoplatonism (it is sufficient to recall Plotinus's motto, "Let us flee, then, to the beloved Fatherland"), while conversion-transformation corresponds to Christianity; for Foucault the Hellenistic model belongs to the pole of return.

As the last "internal characteristic" of the practice of the self, we will consider the *dialectic of universality-exclusivity*. The universality of the practice of the self as an anthropological phenomenon means the universal, all-human character of its declared goal: the task of becoming one's "true self" is posed as the task and vocation of man as such without any historical, social, or other exceptions. But at the same time only a very few people cultivate this practice; its real practitioners constitute a narrow circle. What we have here is an aporia or contradiction in the structure of the phenomenon, and Foucault defines it clearly: "In general every individual is ... capable ... of exercising the practice of the self. No individual is disqualified a priori. ... But on the other hand ... very few people are actually capable of occupying themselves with the practice of the self" (138R; 115F). Having identified the aporia, he notes that there is manifested in it a specific paradigm, a structural form that combines "two elements, the universality of the summons and the exclusivity of salvation," and he asserts the extraordinary importance of this form, its foundational role in the fate of the whole culture and civilization of the West. Indeed, compressed formulas of this paradigm are included among the symbolic precepts of the main cultural formations, going from one to the other: first expressed in the Orphic rites of initiation, this formula is then assimilated by Plato ("There are many Thyrsus bearers, but few Bacchantes": *Phaedo* 69c) and reproduced in the New Testament ("many are called, but few chosen": Matt. 20:16). According to Foucault, "we recognize here a great form [that of conversion] addressed to all and heard only by the very few, the great form of the universal summons which will assure the salvation only of the few" (139R; 116F). As he affirms, "we will find this form anew in the very heart of Christianity," and it is "foundational for our culture" (140R; 116F).

But here is the remarkable thing: even though Foucault ascribes such significance to this form, he does little in the way of achieving a philosophical and anthropological understanding of it. Why is it that all except a few are capable of following the call to fulfill the laws and norms of society, but that no one except again a few is capable of following the call to fulfill the practice of the self which, too, is addressed to all? Here is Foucault's explanation: "Lack of courage, lack of force, lack of endurance, the inability to grasp the importance of this task and to bring it to a conclusion: that is the fate of the majority" (138R; 115F). As we see, he stays on the level of the life-wisdom or rather on the level of his sources, of Hellenistic-Roman thought of the first to second centuries, although he usually does not at all consider this to be his duty, but constantly goes out of the world and the discourse of his sources into a broader context and discourse. It must be supposed that the reason for this is that in this case Foucault would have had to go out into a sphere that he generally avoided: into the sphere of ontology. The "paradigm of universality-exclusivity" receives a transparent interpretation precisely in ontological discourse. If the *telos* of practice is separated from empirical being by an ontological rupture, the practice of the self then turns out to be essentially different from all the practices of ordinary human existence; it turns out to be a radical alternative to the very mode of empirical existence, and needless to say this is a very significant barrier on the path of choosing and following it. But this explanation does not cover all practices of the self. Returning to the above-mentioned bifurcation, the separation of the conversion to the self and the practices of the self into two kinds, we see that the exclusivity of the practice of the self, its alternativeness, its separation by a barrier from all the practices of ordinary existence, occurs if this practice brings forth a cardinal rupture with the initial subject, that is, if it is a trans-subjectivation and presupposes conversion as transformation. But it is possible to go even further: a cardinal rupture can exist even when the conversion to the self is a return to the self. We find this in Platonism and Neoplatonism: yes, "the flight to the beloved Fatherland" is undoubtedly a return — but when we set off on this flight, we discover that we are separated from the Fatherland in the most cardinal way, that is, in a totally ontological way. Therefore the corresponding practices of the self, too, are exclusive and alternative in relation to the practices of ordinary existence. As for the "Hellenistic formation," which for Foucault includes Stoic, Cynic, and Epicurean practices of the first to second centuries, he underscores with particular insistence that these practices were in all respects limited by the horizon of empirical existence. By the same token the "distance" to be overcome by the subject was, in the essential sense, smaller and less significant here; and accordingly the barrier separating them from other anthropological practices was lower and less significant. History confirms this: the practices described by Foucault were fairly widespread and even popular and fashionable. Nevertheless their goal *(telos)* remained special and unique, and they did not merge with other practices, but preserved a stamp of otherness and were the domain of an elite minority. The barrier did not disappear wholly and was still sufficient for Foucault to see in these practices a contrasting combination of universality and exclusivity. However, this contrast is at its purest and most drastic not in them.

* * *

As is generally known, alongside "the practice of the self," there is another major concept of Foucault's conception: "the care of the self." This concept derives from a different discourse and is of a different nature. The practice of the self is an anthropological and philosophical concept introduced by Foucault himself, but the care of the self is a historical concept taken from the arsenal of the culture of antiquity (primarily of late antiquity); it is a translation of the Greek *epimeleia heautou*. Of course in Foucault's conception the care of the self has many additional aspects, dimensions, and connections, becoming to a certain extent also a modern anthropological concept; but this modernization of the concept is not tantamount to its replacement by something else, and when he uses it, Foucault is usually careful to adhere to its historical basis.[16] Sharing with the practices of the self the role of a fundamental concept of Foucault's theory, the care of the self also emerges as a topos of his theory, as a generative center of its concepts; and in accordance with its historical nature, the concepts and principles gathered in this topos also have for the most part a historical character, belonging as they do to the world of ancient culture. However, my analysis pursues mostly anthropological goals. Therefore I will not follow Foucault in his examination of the ancient sources and the description of the care of the self, and its whole topos will be of the most cursory kind.

The 1982 lecture course opens with a methodical and detailed development of the concept of the care of the self. According to Foucault the history of this concept encompasses a millennium, from the fifth century BCE to the fourth to fifth centuries CE, that is, "from the first forms of philosophical activity as developed by the Greeks to the first forms of Christian asceticism" (24R; 13F). The content of the concept changed continuously in the course of this history; however, its evolution always preserved its nature and core meaning. The nature of the concept is profoundly synthetic: "in the concept of *epimeleia heautou* you have a whole corpus defining a manner of being, forms of reflection, practices" (24R; 13F). But primarily, in its very core, this is a specific orientation, "an orientation with respect to the self, with respect to others, with respect to the world." Into its core meaning there enter, first of all, two features of this orientation: first, it demands that "one avert one's gaze from the external . . . and direct it at oneself"; it "presupposes some means of tracking what one is thinking and what is happening in one's thought"; secondly, it "always implies certain actions . . . performed on oneself by means of which . . . one changes oneself, purifies oneself, and transforms and transfigures oneself" (23R; 12-13F). The second

---

16. However, at rare times, in his historical surveys, he allows himself to speak of "the care of the self" in a generalized sense, applying it to epochs outside the millennium in which he places its real historical forms.

feature establishes a connection between the two principal concepts: *the care of the self is realized in the practice of the self.* And, as we already see, the two principles complement each other: the practice of the self can be considered the anthropological core of the care of the self, while the care of the self extends the practice of the self into an integral cultural and social strategy.

The care of the self has another defining and constitutive connection: this is its connection with knowledge of the self, *gnōthi seauton,* the famous injunction of the Delphic Oracle. It is already clear from the aforesaid that the orientation of knowledge of the self is included in the orientation of the care of the self insofar as the latter demands that "one direct one's gaze at oneself" and track one's thoughts and psychical states. But in the course of history the role and the place of the knowledge of the self in the sphere of the care of the self went through profound changes. The cornerstone of Foucault's whole late theory is the thesis that it is precisely the interrelation of two principles, knowledge of the self and care of the self, which is the key characteristic defining the state of man, society, and culture in every epoch.[17] Accordingly, when he analyzes each of the epochs, Foucault always begins by clarifying and investigating this interrelation. The thesis is complemented by a broad survey and assessment: in the course of the history of Western thought and culture there occurred an unfounded and unfruitful rejection of the principle of the care of the self, whereas the principle of knowledge of the self, having been transformed into the principle of knowledge as such and having lost its connection with the care of the self, eclipsed and ousted the latter and became the monopolistically dominant principle of Western thought. The motive factor behind this negative process (Foucault designates, somewhat conditionally, this factor as the "Cartesian moment") consists in the gradual establishment and triumph of the principle according to which "knowledge and only knowledge opens up access to the truth," and meanwhile no (self-)transformation of the subject is required for such access. Going far outside the frame of the problem of the formation of the concept of the care of the self, these propositions adumbrate a new view of the entire path of development of Western philosophy — a view close in part to certain critical receptions of

---

17. Cf. for example: ". . . a dynamic interaction and interplay of *gnōthi seauton* and *epimeleia heautou* . . . can be encountered during the entire course of the history of Greek, Hellenistic, and Roman thought . . . with different specific weights, with different distributions of the elements of knowledge of the self and care of the self. . . . But it is precisely this interweaving which, in my opinion, is very important" (p. 85R; pp. 67-68F).

this path (Foucault traced his own position back to Heidegger[18]), but in part unquestionably original, first of all because the critique is based on positions of the anthropological principle of practices of the self and the Hellenistic principle of the care of the self. I will return to the historico-philosophical implications of Foucault's late theory when I discuss its general contours and vectors.

Representing Plato's *Alcibiades* as "the first theoretical attempt at understanding" the care of the self, Foucault indicates that this attempt at an understanding — and especially its result, the creation of the concept (which, as he emphasizes, the Platonic stage had not yet attained) — must distinguish in the concept, as he says, "its two folds or sections," i.e., "care" and "the self," methodically taking them apart and attempting to understand them separately. As we know, the second section of the concept is for Foucault even more important than the first: he repeatedly characterized the whole plan of his late period as a "history of the subject" and he said that it was precisely the subject that was "the goal of his investigations." When we look at his work from this point of view, we find that the Foucault of the late period left a number of rich though very scattered and unintegrated essays — the beginnings of an integral and original "subjectology." By no means am I trying to reconstruct this subjectology; my only goal now is to pose one of its particular questions: *How, in accordance with Foucault, should one understand "the self" in the term "the care of the self"*? There are a great many scattered remarks about this in his texts, but I think that a fairly complete answer can be found in the following brief discussion of the appropriate passage in *Alcibiades* (129ff.):

> Socrates waits for the answer not to the question . . . what is a man? The question posed by Socrates is much more precise, difficult, and interesting. It is: you must occupy yourself with yourself, but what is this "self" itself *(auto to auto)* . . . ? The question, therefore, bears not on the nature of man but on that which we — we at the present time, since this word does not appear in the Greek text — will call the subject. What is this subject? What is this point toward which the thinking must be addressed as the reflexive and reflected activity, which turns the man back to himself? What is this self? That is the first question. (53-54R; 39-40F)

---

18. Cf. for example: "What is the relation of the subject to truth? What is the subject of truth, what is the subject that speaks the truth? . . . [I]t is on the basis of Heidegger that I am trying to reflect on all these things" (p. 214R; p. 182F).

PRACTICES OF THE SELF AND SPIRITUAL PRACTICES

This is an extraordinarily rich text, in which — as often happens with genuine philosophers — into the form of a question is incorporated a content containing not only some answer but a whole series of new questions and answers — or in other words a content that provides a trampoline for a jump into a certain topos. In the present case this is the topos of the "self." In Foucault's text there is imbedded a series of identifications that not only propel us toward the answer of the question but also reveal to us the usage rules of the words employed by Foucault. Here is this series:

"the self" (le soi) = the self itself (auto to auto) in the Platonic discourse = "the subject" = "the point toward which the thinking as the reflexive and reflected activity, which turns the man back to himself, must be addressed"

The last item in this series of identifications is not an ordinary term or locution but a true unfolded definition. This item lends itself to prolonged investigation; in its richness of content it is capable of becoming the nucleus of a comprehensive essay in subjectology. One can discover in it more than a few connotations referring to Hegel's *Phenomenology of the Spirit* (for Foucault one of the most important texts of Western thought), to Descartes, to . . . At present, however, I will only mention the two most crucial aspects. The first touches on that which is not present, but absent, in the series; there is an obvious lacuna in the set of identifications of the subject: there is no "I" here, no Ego. In contrast with Descartes, Foucault's subject is identified not with "I" but with "the self"; corresponding to it is not a personal but a reflexive pronoun (this is the case both in the French and in the Russian). This linguistic fact is of fundamental importance: it immediately takes us into the nature of this special subject.

From the point of view of subjectology there is an enormous difference between "I" and "the self."[19] The subject-"I," acting from the first per-

---

[19]. Soon after Foucault's death this difference was thoroughly analyzed by Paul Ricoeur with reference to Foucault's work on the conceptions of the practice of the self and the care of the self. Ricoeur, however, examined this difference from a point of view quite distinct from the one taken by Foucault and by myself. He wrote: "the primacy of the reflexive mediation of the immediate position of the subject as this position is expressed in the first person singular: 'I think,' 'I exist.'" Paul Ricoeur, *Ya-sam kak drugoi* (Oneself as Another) (Moscow: Izdatelstvo Gumanitarnoi Literatury, 2008), p. 15. (French title: *Soi-même comme un autre*.) Of course in both Foucault's project and Ricoeur's study the difference between "I" and "the self" is exploited with the aim of revising Descartes' subjectology, but the strategies of revision turn out to be very different.

son, is the sovereign active center, the generative source of all of a human being's activities. As is known, a concept like this which arose as a result of the solution of the problem of individuation did not exist in antiquity; the problem of individuation had not yet been posed and it certainly had not yet been solved.[20] But Foucault supposes that the concept of "the self" already existed, at least in its rudimentary foundations. It is precisely with this concept that he identifies Plato's famous "self itself," and it is also with this concept that he connects his "subject." What is the character of this concept? As grammar teaches, a reflexive pronoun "turns an action back toward the actor himself"; reflexivity is a return to the self, toward the self, attraction toward the self, possessiveness, so to speak; in other words, a reflexive pronoun, categorically not being a possessive pronoun,[21] nevertheless has a sense analogous to possessiveness. In its meaning, "self" is a kind of "possessive correlative" of "I." Thus, if one follows the leading thread of the language, the subject-"self" is the subject actualizing the return of every action toward the self, and by virtue of this it can be thought as the center of *attraction (or possessiveness)* into which all of a human being's activities are drawn and by which they are organized. But generally speaking it is not supposed that they are generated by this center; the "subject-self" is rather their distributor than their generative source. Accordingly, in the aspect of thinking, the "subject-self" is only the point "toward which thinking must be addressed"; and generally speaking it does not have to be the point in which thinking is engendered and from which it emanates. Even if it is the center or locus of thought, it is only a locus where thought must be brought and installed by means of conversion to the self; it is not a locus that is the generative source of thought.

    These properties of Foucault's subject give us grounds for a fully definite conclusion: of course this "subject," which is not the First person, I, *cogito*, but is only the "possessive (or attractive) self," is also not the subject in the proper philosophical sense; and when Foucault uses this term in his investigations of the ancient practices of the self, it must be understood in a conditional sense. We can find in Foucault a direct statement concerning this: "Not a single Greek thinker ever found or sought a definition of the subject; I will say straight out that *there was no subject*.... Classical antiquity did not problema-

---

    20. I have presented an examination of the genesis of the subject in the light of the problem of individuation in chapter 3 of *Diogenes' Lantern* (on Descartes).
    21. The Russian term for possessive pronoun *(pritiazhatel'noe mestoimenie)* literally means "attractive pronoun," and so I have rendered the noun *pritiazhenie* either as attraction or possessiveness, depending on context. — Trans.

tize the constitution of the self as subject."²² Besides this negative assertion, one can also find characterizations of the positive meaning and content of "subject constructions" that give at least a partial answer to the question, *But if there is no subject here, who or what is there?* Cf. for example: "I think that there does not exist any sovereign subject, any founding *(fondateur)* subject, any universal form of the subject which could be encountered everywhere. I am extremely skeptical about and hostile to such a conception of the subject. I think, on the contrary, that the subject is constituted through practices of subjectivation *(assujettissement)* or, in a more autonomous fashion, through practices of liberation or freedom . . . starting, of course, from a certain set of rules, styles, and conventions found in the cultural environment."²³ Evidently, that subject construction or formation which Foucault associates with Plato's discourse is also a certain form of subjectivation. A "possessive pro-noun," or "attracting point," would be a perfectly appropriate name for this construction if there were not an ambiguity here: the possessive pronoun is not only *soi*, but also *moi* and *toi*. Therefore to the pre-personal subject formation described by Foucault as corresponding to the "subject" of antiquity we will give the name: "possessive (or attractive) self."

"Subjectivation but no subject" — this character of the subjectology of Foucault of the late period was tirelessly asserted and advocated also by Deleuze, who after Foucault's death wrote and spoke a great deal about his work. He repeatedly stated that Foucault "never returned to the subject," that in Foucault one found "subjectivation without a subject," "a subject without identity," "an ensemble of impersonal forces," etc., and finally that "Foucault describes subjectivation as a process and 'I' as a relation (a relation to one's I)."²⁴ However, subjectivation is a rather indefinite and fuzzy notion, and Deleuze did his utmost to convince us: in Foucault's discourse, subjectivation is extremely far from the classical subject; it does not retain any of the subject's predicates and does not have any personal characteristics. "Subjectivation is a production of modes of existence or styles of life . . . that which Nietzsche called the invention of new possibilities of life."²⁵ Underscoring to the extreme the anti-subject and impersonal nature of these modes, Deleuze arrives at a certain hypertrophy, where he stylizes Foucault's thought in the spirit of his own passion for dehumanization and

22. Foucault, "Le retour de la morale," *DE II*, no. 354, p. 1525. (Italics added.)
23. Foucault, "Une esthétique de l'existence," *DE II*, no. 357, p. 1552.
24. Gilles Deleuze, *Peregovory* (Negotiations) (St. Petersburg: Nauka, 2004), p. 124. In French: *Pourparler 1972-1990* (Paris: Les Éditions de Minuit, 1990).
25. Deleuze, *Negotiations*, p. 151, p. 155.

the reduction of anthropological reality to topology and physics: "Subjectivation had little in common with the idea of the subject. It is rather a question of an electric or magnetic field... of fields of individuals."[26] Things clearly do not add up here: when in place of the subject we find an electric or magnetic field — that's Deleuze, not Foucault.

Returning to Foucault's "subject," let me note yet another one of its properties that is important for this analysis. The discourse of practices of the self and of the care of the self is an active discourse, a "verb" discourse; it speaks expressly about how "one should occupy oneself with oneself," about an individual's actions, and it does not operate with any abstract concepts, with any essences. Therefore the protagonist of this discourse, the "subject-self" or "the point toward which the thinking must be addressed," is characterized and defined exclusively by activity as such directed toward him; and strictly speaking one cannot say that this activity "fills" him as if he were a kind of receptacle with some substantial or essential "contents." In this sense he is not a "subject of activity" but a "subject-activity," a "point-activity" (it would perhaps be better to speak of "actions," thus avoiding confusion with the "methodology of activity"[27]). Like a "point-activity," which has only activity itself or actions as its sole content and definition, it is a dynamic formation and its nature is one of action. And consequently it cannot receive adequate conceptualization in an essentialist discourse. This is a characteristic of personology or subjectology not only in the discourse of practices of the self but also in the consistent discourse of all practices in general, and in my opinion it is one of the main factors leading to the specific reception of Aristotle in Foucault: fixing his gaze on ancient culture and thought in the context of practices of the self, Foucault declares that "Aristotle is not the peak of antiquity, but an exception" (adding as a challenge: "as everyone knows") (30R; 19F).

It is clear that the "self" in the formula "care of the self" is not yet the "true self" toward which one must progress by means of the practice of the self which embodies the principle of the care of the self. Therefore both of the above-examined properties of Foucault's subject which reveal it as a "self" having only a "possessive" but not a personal nature, and as a "point-activity" — both of these properties characterize it *before* it constitutes itself or, as Foucault prefers to say, before it "establishes itself," *s'établit*, in the

---

26. Deleuze, *Negotiations*, p. 125.

27. The methodology of activity, founded by Georgy Shchedrovitsky, is an influential philosophical school in Russia in the second half of the twentieth century.

practice of the self. One can expect that they will manifest themselves in some way and have an effect on its constitution. I will return to this problematic of the constitution of the "subject" in Foucault more than once, and in conclusion, in Section II, I will compare it with the constitution of the person in spiritual practice.

According to Foucault the concept of the care of the self "supported an extremely rich and dense ensemble of concepts" (24R; 13F). But by virtue of the historical character of the concept this ensemble contained very few universal concepts that were connected with care of the self as such, and not with one of its particular forms. We will confine ourselves to mentioning two of them, the most essential ones: "the art of living," *technē tou biou*, and *chrēsis*, "use," the universality of which is underscored by Foucault: "We will encounter the concept of *chrēsis* in the course of the whole history of the care of the self" (73R; 56F). The theme of "the art of living," like many other themes, unites Foucault with Pierre Hadot, who during those same years wrote about "the conception of philosophy as an art of living, as a form of life," and for him this conception was deeply rooted in antiquity. Analogously with Hadot, Foucault declares that "the art of living (this famous *technē tou biou*) ... beginning with Plato ... will become the principal definition of philosophy" (103R; 84F). But, in contrast with Hadot, Foucault demonstrates the connection of this art with the principle of the care of self, and he finds that with the flowering of the care of self in the Hellenistic epoch two principles coincide: "the art of existence ... and the care of the self, or more concisely the art of existence and the art of the self, coincide more and more manifestly.... [T]his care of the self, having arisen in the context ... of the development of a *technē* (an art of living) occupied the whole place defined by the *technē tou biou*. That which the Greeks sought ... for so many centuries, from the beginning of the classical age, this *technē tou biou* is now ... entirely occupied by the principle that one must take care of oneself" (201, 525R; 171, 465F).

As for *chrēsis*, it is the case, as Foucault underscores, that in the context of the care of the self it manifests itself not in the usual sense of simple use but rather as a certain obligatory form of use, as a concept defining a special mode, the mode of adequate relation to the things of the surrounding world. This mode that is dictated by the care of the self is connected with the subject nature of the "self"; expressing a "singular and transcendent position of the subject in relation to that which surrounds it" (73R; 56F), it becomes one of the essential internal characteristics of the "true self." Foucault does not trace the history of *chrēsis* in Christianity, but it can be acknowledged that this category was assimilated by early Christian culture with almost the same meaning and content as in antiquity and became part of the arsenal of patristics and ascetisicm. Let us also note that the concept of the *"culture of the self,"* which Foucault uses widely and describes in very general culture-philosophical categories, is nonetheless not universal: according to Foucault, it is possible to speak of "the development of a certain culture of the self only beginning with the Hellenistic period"; but this culture had not yet been formed in classical antiquity. "Culture of the self" is not an ancient, but a modern concept, a generalization and expansion of the "care of the self" introduced by Foucault. This is a mode of existence in which "the relations of oneself to oneself acquire a particular intensity and particular value";[28] and here this mode, like the practice of the self, is characterized by a contrasting combination of the universality of its goals and values with exclusivity, with the narrowness of the circle of those who are capable of really following this mode. According to Foucault, the culture of the self enters into the composition of "the cultural form known as *paideia,*" and it very closely resembles the German Enlightenment concept *Selbstbildung,* "formation of the self."

Finally, the fundamental vocabulary of Foucault's conception also indisputably contains *ascesis (askesis),* or "philosophical ascesis." In modern culture this term is usually associated with the Christian church and the institution of monasticism, but it also has a rich history in antiquity dating back to Pythagorism. Needless to say, it is with reference to this history that Foucault uses this concept. He accepts the usual definitions of ascesis ("a diligent and zealous training," a practice, an ensemble of practices, "an exercise in self-making," etc.), and then he puts forward and develops a number of aspects of the concept that he finds necessary. First of all, for Foucault ascesis closely resembles the practice of the self, but it is presupposed to be a somewhat broader and more general concept, as can be seen, for example, from this characterization of it: "It seems to me that in pagan ascesis, in philosophical ascesis . . . the point is to unite oneself with oneself by means of a specific practice of the self. . . . To make the truth one's own, to become a subject of true speech: that is . . . the very essence of this philosophical ascesis" (360R; 317F). As is clear from this definition, ascesis shares with the practice of the self the role of a constituting activity: "Ascesis . . . constitutes, and it is its role to constitute, the subject as subject of veridiction" (402R; 355F). The synonymous term "philosophical ascesis" is used in Foucault's

---

28. Michel Foucault, *Histoire de la sexualité,* Vol. 3: *Le souci de soi* (Paris: Seuil/Gallimard, 1984), p. 57.

discourse to accentuate two aspects: first, the very close and mutual connection of ascesis and philosophy in the Hellenistic epoch; and, second, the difference between pagan and Christian ascesis.[29] Foucault sharply distinguishes between these two formations of asceticism, and we will touch on their differences later on.

I.2. Sketch of the Conception

So far I have described the conceptual apparatus of Foucault's theory. The next step is to describe what this apparatus was intended for, which means:

> to describe the general structure and composition of what was accomplished by the thinker;
>
> to reconstruct, as far as possible, his incompletely realized projects;
>
> and, summing up, to present the full contours of Foucault's entire project, the work of the last years of his life. Finally, one has to evaluate this project in the context of the philosophical and cultural situation of our time.

Two formulas are always used in the attempts to convey briefly the main theme and essence of Foucault's last project: "history of the subject" and "practices of the self." In describing the language of the project, we naturally placed at the center the "practices of the self," a new fundamental concept invented by Foucault. But when characterizing his project the author himself more often emphasized another formula. "I have conceived the history of the subject," he said in his last interview; and in the introduction to his 1982 lecture course this is how he posed its fundamental question: "In what form of history are relations between the 'subject' and 'truth' established in the West?" (14R; 4F). Of course, there is no conflict between the two formulas; there is only a simple logical connection: according to Foucault, it is in practices of the self that the subject is constituted, so that the "history of the subject" is nothing other than the sum of practices of the self realized by the subject. *Practices of the self are a new way of looking at the constitution of the subject, a way of looking that engenders a new approach to constructing the history of the*

---

29. For a discussion of the Christian concept of ascesis, see Sergey S. Horujy, *K fenomenologii askezy* (Toward a Phenomenology of Ascesis) (Moscow: Izd. Gumanitarnoj Literatury, 1998), pp. 72–82.

subject. Foucault blazed an original trail in contemporary philosophy that is defined by him as "an attempt to situate the subject in the historical domain of practices and processes where it has been continuously transformed. This is precisely . . . the path I have chosen" (572R; 506F). This is what the chosen path dictated: "[I]t is necessary to start with the history of practices which serve as the foundation for the forms of reflexivity that constitute the subject" (502R; 444F). And the philosopher begins his last large project with a major reconstruction of the history of practices of the self.

Foucault describes the large structure of this history by means of his schema of three major formations (forms, types, or models) of practices of the self that occurred successively in the history of the West. "On the level of practices of the self there are three great models which followed one another in historical succession. The model which I call 'Platonic' model and which gravitates around recollection. The 'Hellenistic' model, which is centered on the self-finalization of the relation to the self (*l'autofinalisation du rapport à soi*). And the Christian model, which is centered on self-exegesis and self-renunciation" (284R; 247F). Below I will characterize these three models in a cursory fashion, but it can be said at the start that, as a whole, Foucault's schema, as well as his entire last project, bears the stamp of incompleteness. We find a relatively clear picture only of the millennium that Foucault allots for the history of the "care of the self" (from the fourth century BCE to the fourth-fifth centuries CE) and that concludes with the establishment of the foundations of Christian asceticism, where, according to Foucault, we find in a pure form the "Christian model" of the practice of the self. How this schema was planned to be continued is something we can learn only from some of Foucault's scattered remarks, although it was part of his project to give it a very clear and complete continuation; cf. "What I want . . . is to attempt to situate inside a historical field *articulated as precisely as possible* the ensemble of those practices of the subject which have been developed from the time of the Hellenistic and Roman epoch *up to the present time*" (213R; 181F; italics added). When discussing the "Christian model," I will cite Foucault's principal statements about how he envisaged the further history of the practices of the self.

I.2.1. Genesis of the Practices of the Self: The Platonic Model

The Platonic model of the practices of the self, corresponding to classical antiquity, is reconstructed by Foucault in an original way — by analysis of a single text, Plato's *Alcibiades* (or more precisely the *First Alcibiades*), and by discussion of a single hero, Socrates. (*Voluntaristic selection* is in general one of the main tools of his methodology in the creation of the conception of practices of the self.) He considers this text and this figure to be extraordinarily significant: Socrates is "the teacher of the care of the self" and *Alcibiades* represents "the first theory and even . . . the sole global theory of the care of the self" (61R; 46F).

In Foucault's conception the key characteristic of every formation of the care of the self is the correlation between this principle and the principle of self-knowledge. In the present case the two principles are symbolically united in the figure of Socrates: he teaches both the one and the other and introduces into philosophy and Greek consciousness both the principle of self-knowledge and the "art of the self." Accordingly, for his pupil, "for Plato . . . all care of the self . . . becomes a form of knowledge and self-knowledge" (62R; 49F), just as in general at this stage care of the self is essentially self-knowledge: "The major if not the exclusive form of care of the self is self-knowledge: to occupy oneself with oneself is to know oneself" (98R; 80F), so that as a result "the whole surface of the care of the self is occupied by the imperative of self-knowledge" (281R; 244F).

The fact that this stage has the character of a beginning is manifested in a whole series of features. The anthropological paradigm was just being formed; care of the self, as we have said, had not yet crystallized into a separate concept. Many aspects of the later concept were still absent in the Greek consciousness. Socrates presented care of the self to Alcibiades as only a means; he presented it not as a goal in itself but only as necessary preparation, as training for Alcibiades' future activity in the role of ruler; in other words, it was required *not for the purpose of being a human being but for the purpose of being a ruler.* This means, it lacked the aspect of universality, of belonging to all humans, that which above I have noted to be the predicate of the universality of the practice of the self. Indeed, its connection with the practices of the self was not understood yet: the discourse of the care of the self does not send one back to the fully developed sphere of practices of the self that was created in Greece by Pythagorism and the mystery cults and that in effect was the archaic prototype of the care of the self (although, as Foucault points out, Plato's philosophy has "a multitude of traces of such technologies," but they will be gathered into the sphere of care of the self only in a later epoch). Digressing from his analysis of *Alcibiades*, Foucault lists the basic types of these "archaic technologies of the self." They include: 1) rites of purification, into which, according to tradition, Pythagoras himself introduced such a highly organized intellectual component as the "examination of consciousness," a regular review and assessment of all one's acts and thoughts; 2) "technologies of the concentration of the soul" *(pneuma)* and the protection of its integrity, self-gathering, and self-concentration; 3) "the technology of renunciation . . . *anachōrēsis*, withdrawal into the self . . . this is a certain method of leaving the world, of being absent, without giving up one's place in it . . . of breaking all external connections, of not perceiving and not noticing that which is going on around one" (63R; 49F); 4) the technology of testing oneself, of testing one's ability to overcome temptations, to endure bodily and spiritual suffering, etc. Later on all of these types of technologies will become the basis of the repertoire of the practices of the self of the mature "Hellenistic model" (and also to a large extent of the Christian model).

By contrast, other aspects of the care of the self were already clearly present. From the very beginning an important role in the care of the self was played by the intersubjective or transindividual aspect, the condition of the necessity of a mentor. The mentor was included organically in the original form of the care of the self, so that there was not even any need to assert the necessity of having one. The fact of the matter was that, at this stage, the subject of the care of the self was supposed to be a young man who was just starting out in life and who could get all his ideas about the nature of the care of the self and how it should be cultivated from his mentor only. Therefore "in the early Socratic-Platonic form the care of the self was first of all that which a young man needed in his relations with his teacher, or with his lover, or with his teacher and lover together" (53R; 39F). Without going deep into this theme, Foucault nonetheless gives us to understand by a series of cursory remarks that in his conception the full and "habitual" variant of this "early form" of the care of the self is the last one mentioned, that is, the one in which the mentor is both a teacher and a lover.[30] I have already noted above that, at this stage, the activity of the mentor or psychagogue is still within the sphere of pedagogy. The last feature constitutes a supplementary addition: according to Foucault this is pedagogy combined with pederasty, and the Other in the paradigm of the care of the self (the practice of the self) is the teacher-pederast.

Finally the cardinal question regarding the "second fold" of the concept, that "self" at which the care must be directed, also has a fairly clear and

---

30. Cf. for example: "The practice of the self . . . is inscribed in a special loving relation between teacher and pupil" (p. 232R; p. 197F).

definite answer at this initial stage (paradoxically, in the next Great Formation, the Hellenistic one, this will become much less clear!). This question is posed directly in *Alcibiades,* and here is the answer that is obtained, in Foucault's paraphrase: "What is this self with which one must occupy oneself? — It is the soul" (69R; 53F). Foucault shows that the Platonic treatment of the "soul" corresponds to the above-described conception of "subject-self" and "subject-activity" (but not "subject-*cogito*"). But the paradigm of care of the self/practice of the self distinguishes between the "self" and the "true self" acquired through the fulfillment of the care/practice. In the Platonic form the fulfillment of the care of the self essentially coincides with the knowledge of the self; and by its nature this knowledge is anamnesis, recollection. In Foucault's treatment anamnesis is exactly the form in which care of the self and knowledge of the self coincide: "The soul discovers what it is by recollecting what it has seen. And it regains access to what it has seen by recollecting what it is" (281R; 244F). Therefore, Foucault defines the quintessence of Platonic subjectness and the Platonic formation of the practice of the self as a "model of the subject's recollection of itself."

I.2.2. The Hellenistic, or "Ethical," Model

The "Hellenistic model" of the practice of the self is the chief object of Foucault's attention and interest, and perhaps even the raison d'être of all his late theories. He spends more and more time on this model, distinguishes it as special in the whole history of the anthropological paradigms studied by him, repeatedly calls it and its epoch the "golden age of care of the self," the "highest point of evolution," and so on. He ascribes to it not only a historical value but also a value for modernity; it becomes his personal preference and passion: As his sensitive interlocutor underscores, "for M. Foucault, as well as for me, all this was not only an object of historical interest.... This was precisely the way he conceived philosophy at the end of his life."[31] But on the other hand he allots to this model a comparatively short period, the first to second centuries, and even within the limits of this period he carries out his voluntaristic selection, leaving out several major phenomena that are directly related to the object of his studies (first of all, gnosticism and skepticism). He also pays no attention to the most significant practices of the self in the ancient world, the ones developed in the mystery cults and in

---

31. Hadot, *Spiritual Exercises,* p. 341.

Neoplatonism. Therefore, in spite of the impressive profundity of Foucault's studies, his model raises more than a few doubts and questions. What does it really tell us about man? In this model, has man not become the servant of Foucault's sympathies and passions? To what extent is this model useful for any general anthropological and philosophical conclusions? We will need to return to these questions when we sum up Foucault's project as a whole.

The model is constructed in historical logic: first Foucault pinpoints the changes that occurred in the sphere of the care of the self and the practice of the self compared with the epoch of Socrates and Plato. He singles out three main changes, all of which had an analogous character in that they augmented the significance of the care of the self and extended the sphere of its application. In the classical epoch, care of the self was subordinate to three conditions or limitations: first, it was not viewed as a general anthropological principle, but was regarded as necessary only for a specific group, the citizens of the polis; secondly, it was regarded not as a goal but only as a means of preparing a young man for the role of ruler; and thirdly its realization was limited to knowledge of the self. In the Hellenistic epoch all three of these limitations are removed and disappear, which imparts to the phenomenon of the care of the self a fundamentally new nature and status. The phenomenon "acquires the scope of a veritable culture of the self" (231R; 197F).

When the first limitation was removed, "the care of the self became a general and absolute principle, an imperative imposed on everybody, all the time and regardless of status" (99R; 80F). In other words the imperative of care of the self acquired the predicate of universality, which Foucault calls "expansion" and "generalization of the imperative." When the second limitation was removed, care of the self and practice of the self became autonomous and acquired a purely anthropological nature. They were no longer subordinate to any external, extraneous goals; from this moment on, the care of the self, the development of a particular relation to oneself, is viewed as a goal in itself. The "self," the "subject-self," is clearly defined as the *telos* of the practice of the self: "in the practice of the self, as it emerged and formulated itself in the last centuries of the so-called pagan era and in the first centuries of the Christian era, the self *(le soi)* basically emerged as the goal, the end point" (276R; 239-40F). The practice itself thus becomes a completely independent anthropological practice, not integrated into or subordinate to any practices of power, any social or other practices. The removal of the final limitation is connected with a cardinal expansion of the sphere of the care of the self and the practice of the self. Into this sphere there return many elements and features of the ancient practices, Pythagorean and other, that were never limited to practices of power, and the content of this sphere thus becomes very diverse and heterogeneous. The practice of the self approaches holistic practice, which requires the participation of all the levels of man's being. The principle of self-knowledge is by no means rejected here and is not removed from the practice of the self, but now this principle forms just one part of this practice; as Foucault says, "the form of self-knowledge ... is integrated into a vaster ensemble" (100R; 81F). At the same time this principle is transformed, acquiring a highly interesting development (which we will examine in our discussion of the paradigm of conversion).

The transindividual dimensions of the practice are also transformed. The rule of the necessity of the Other, the Mentor, remains unchanged, but the role of the Mentor and the forms of relations with him change in a cardinal way. Foucault conveys the fundamental content of the changes by the formula "separation from pedagogy." This formula expresses not only the obvious fact that when everyone has become (a priori) the subject of the care of the self and the formation of the "true self" has become the task of the practice, mentorship in such practice can then evidently no longer be a part or variation of simple pedagogy. This formula also has a positive content: in the new stage the Other must be a "specialist in adjusting the relations of the subject with himself"; he must be a possessor of that truth about the subject the revelation of which is the goal of the practice of the subject. The mission of the Other changes cardinally: it becomes more profound and is interiorized. He must gain knowledge of and penetrate into the very core, so to speak, of the "subject-self"; and he must indicate in what manner this "self" should be transformed. With respect to such a task, all pedagogy turns out to be an overly superficial form, and psychagogy must separate itself from it. In addition, as we have already noted above, it clearly begins to approach what Christianity will formulate as care of the soul, pastorship (we will return later to the distinction between these two psychagogies). However, according to Foucault, the role of the Mentor, the Teacher, in the Hellenistic practices is manifested most precisely and fully through a classical ancient comparison, about which he speaks a great deal: "the comparison between medicine, the art of the helmsman, and spiritual directorship, the government of oneself and of others." This triple comparison bundles together a "whole set of Greek and Roman concepts indicating ... one and the same type of knowing, one and the same type of activity, one and the same type of probable knowledge" (276R; 239F). The last item is especially important for Foucault: it expresses the specificity of a given type of activity as "activity which is simultaneously rational and intuitive," which is governed only by probable or likely arguments and does not have exact data or proofs at its disposal. In order to assimilate this specificity the attempt was made in antiquity "to institute a certain *technē* (an art, a developed system of approaches, general principles, and concepts)"; and as a result the triple comparison in combination with the *technē* that grew out of it constituted, as Foucault says, "the real matrix of a theory of government." It is this matrix that serves as the basis for the psychagogy, for the discourse of the Teacher in the Hellenistic-Roman practice of the self.

There is another aspect of the "discourse of the Teacher" that became an object of special and extensive investigations on the part of Foucault: namely, the necessary property of the Teacher's speech conveyed by the ancient concept *parrhēsia*. After going through a complex history in antiquity, this concept did not survive the latter and was not assimilated by modern languages. The spectrum of its basic meanings is more or less encompassed by the accepted translation, "free speaking"; this translation lacks clarity, however, and we prefer not to use it. As Foucault shows, in the context of the Hellenistic practices of the self, *parrhēsia* is loaded with many additional meanings and connections, thereby acquiring the character of a special "spiritual exercise" and being integrated into the practice as an essential element of the latter. Foucault devotes to it the most attentive study: in addition to detailed analyses of it in the 1982 course, he wholly devotes to it the 1983 and 1984 courses, as well as the workshop in Berkeley. It is thus necessary for us to understand not only the essence of this concept and its role in the practice of the self, but also the reasons behind the philosopher's intense interest in it (after all, the "discourse of the Teacher" does not directly belong to the core of the practice of the self that is the path of the "self").

Foucault has many formulas/definitions that express different aspects of *parrhēsia*. His first, preliminary references to it already underscore its mutual, "binary" character: in the practice of the self *parrhēsia* is not so much a "property of the Teacher's speech" as it is a certain fundamental quality of the Teacher-Pupil *relationship*, of the whole atmosphere of their relations. Yes, the realization of *parrhēsia* is directly a task of the Teacher: "*Parrhēsia* . . . is the principle of the verbal behavior which one must have with the other in the situation of the mentorship" (187R; 158F). But this is a principle whose implementation must establish between the Pupil and the Teacher not just a specific atmosphere of communion but also a specific ethics. "*Parrhēsia* is the opening of the heart; it is the necessity that the two partners not hide anything from each other of what they are thinking and that they speak frankly to each other" (158R; 132F). And, as such, it was recognized as "one of the fundamental *ethical* principles of direction" (158R; 133F; italics added). The ethical content of *parrhēsia* is connected with its total unselfishness and, moreover, with its moral elevatedness: "... one who practices *parrhēsia* ... never has any direct or personal interest in this exercise of *parrhēsia*. The exercise of *parrhēsia* must essentially be commanded by generosity. Generosity with regard to the other individual is at the very heart of the moral obligation of *parrhēsia*" (417R; 369F). This ethical content of *parrhēsia* constitutes its chief distinction from rhetoric, though to a certain degree it has the same goals as the latter (to convince the recipient of the speech, to obtain a definite reaction from him). But more importantly it is here that we find the key theme of Foucault's reception of the Hellenistic practices of the self: for Foucault these practices are, first of all, a sort of laboratory for the creation of a new ethics or morality; and he sees the chief significance of the Hellenistic model and its historical value precisely in the fact that "there was formed in it a certain morality" which had an enormous, though hidden, influence on the subsequent history of the ethical consciousness and which became, as he says, the "matrix" of more rigoristic ethical systems in the future.

But *parrhēsia*, as Foucault underscores, operates in the practice of the self not only as an ethical concept but also as a "technological" one. This is precisely a method of speech that is necessary for this practice in order to achieve its direct goal, the *telos*, i.e., "access to the truth" and the becoming of "oneself," one's "true self." And this "technological" function implies a whole series of requirements that define the nature and structure of *parrhēsia*.

It must be understood, first of all, that the Teacher's speech is not a secondary and auxiliary but the most central and critical element of the practice, that it is the sole source and the "place" of that truth to which the Pupil seeks access. "The truth is wholly present in the teacher's speech and only in it" (394R; 348F). *Teacher-pupil* communication is thus generative in the practice; it is pivotal for the practice, the only source that feeds it. There is another important element here: the Pupil's relative passivity in this communication. According to Foucault the conception of *parrhēsia* presupposes that the Pupil's role consists in being silent: "His [the pupil's] task is, essentially, to remain silent. ... *Parrhēsia* is basically that which responds, on the part of the teacher, to the obligation of silence on the part of the pupil" (394-95R; 348F). As Foucault remarks, "etymologically, *parrhēsia* is the fact of saying everything" (395R; 348F); and in this case this formula can be understood in the strongest sense: everything that is and must be in the practice of the "self"-practicing Pupil must be uttered — as well as communicated to the Pupil! — by the Teacher. And in order to achieve this, he must literally *lay himself out* before the Pupil. His speech must bear a visible stamp of his truthfulness, and to achieve this he must wholly put himself into his speech and assure "the sensible presence of the speaker in what he is saying," where the subject of speech coincides with the subject of behavior. That is precisely the genuine nature of *parrhēsia*: "What authenticates the fact that I am saying the truth to you is that I as the subject of my behavior . . . am absolutely and totally identical to the subject that I am when I am saying to you what I am saying. . . . That is the essence of *parrhēsia*" (441R; 389F).

As this makes clear, in the context of the practice of the self *parrhēsia* turns out to be an extremely interesting phenomenon that leads one out to many further questions and studies. Regretting that we do not have access to the last results of Foucault's studies, let me note two things: First, the "essence of *parrhēsia*" as I have described it is such that it is clearly not just a practice of speech but also a practice of the self or at least one of the "spiritual exercises" (an element of the practice of the self which I will address below). A peculiar structure is revealed: included in the practice of the self performed by the Pupil — and necessary for achieving the *telos* of the practice — are, if not the practice of the self of the Teacher, then at least "some of his spiritual exercises"; in other words, an essential part of the practice is *placed from the Pupil to the Teacher*. Kindred systems of anthropological practices (the spiritual practices that we will discuss in the next section) do not exhibit this kind of structure (for example, the Zen Pupil, when solving *koans* and fulfilling other kinds of tasks, must extract by himself from his communication with the Teacher an experience that gives access to the truth). Here the work of accessing the truth is not placed on the Teacher, and the barrier of access is higher. I am just noting this fact; at this point I will not attempt to interpret it. But let me indicate that, to understand it, one must take into account such factors as the specifically "possessive," but not "personal," nature of the subject of the practice (which increases his passivity), as well as the entirely immanent nature of the *telos* of the practice, which, according to Foucault, wholly situates the "true self" in empirical being and corresponds to self-subjectivation, not trans-subjectivation. In the light of these two factors the practice described by Foucault appears as a *moderated* practice of the self, directed at a *lowered* (purely immanent) *telos*. But this is only a preliminary and hypothetical assessment, and the situation demands further analysis.

Secondly, in the light of Foucault's description of *parrhēsia* we see that the Hellenistic practices of the self were also practices of the development and deepening of the forms of human communion. The condition of the "sensible presence" of the speaker himself in his speech (in other words, the condition that oneself be placed into the communion), the ethics of complete openness in the Teacher-Pupil relationship, compels us to notice that this relationship exhibits incipient elements of a certain type of communion, namely, the "communion of persons," "personal communion," which is new for antiquity but will be developed in Christianity. The paradigm of personal communion, created by Christian culture and directly engendered by the nature of Christian experience as the experience of meeting/union with Christ, God the Person, God and man — this paradigm brought with it a new model of communion as an increasing exchange of existential and personal contents, as a chain reaction of the ever-deepening communalization of "living worlds," of the creation of a common personal medium, of a "bi-unity" (to use a term from Russian religious philosophy). Progress toward this model and some of its features are clearly visible in Foucault's descriptions, in which he sometimes spontaneously goes over to a "discourse of personhood," although he knows that this discourse did not exist in antiquity: "[O]f course, personal relations are necessary here. Correspondence represents personal relations. Even better is personal contact during conversation. And still better is the sharing of life-experiences, of a long chain of life-examples, passed from hand to hand, as it were" (440-41R; 389F). To be sure, Christianity was not an isolated phenomenon, and in some of its other parts the world of late antiquity too was, to one degree or another, big with the "birth of personhood."

The end of the last quotation touches on another question that is closely connected with the practices of the self: the question of the preservation and undistorted transmission of the specific experience of the practice of the self, the experience of "access to the truth." As we will see in Section II, in spiritual practices the problem of such transmission is one of the main problems, and its solution consists in the necessary creation of a special anthropological and historical mechanism, of a *spiritual tradition* whose mission is the identical reproduction and translation of the experience of a given practice. But in the practices described by Foucault this phenomenon does not arise. Even though the "long chain" along which experience is identically transmitted is good for the practice, it is by no means obligatory, and Foucault mentions it only because he had just been talking about the Epicureans, in whose school such a chain had been created and maintained: "In the dynasty of the Epicurean leaders a direct ascent to Epicurus through the transmission of a living example, through personal contact, was indispensable" (421R; 373F). It was precisely this direct experiential succession of the Teacher from the Proto-Teacher that was viewed as guaranteeing the truthfulness of the Teacher's speech; and if the Teacher, following *parrhēsia*, placed himself into his speech, the Pupil was able to receive eventually the speech of the Proto-Teacher himself, Epicurus. This phenomenon, to which Foucault gave the name "vertical translation," represents a certain prototype, one of the embryonic forms of spiritual tradition. We do not find this in the other practices examined by Foucault: Seneca, Epictetus, and Marcus Aurelius were Teachers for whom the problem of translation did not exist. It is natural to connect this circumstance, once again, with the "moderated" character of these practices, the experience of which, in virtue of the immanence of the *telos*, did not differ radically from the usual forms of existential and cultural experience and did not include that specific instrumentarium and canon whose creation and translation are accomplished only by spiritual tradition.

Before examining the system of exercises that constitute the "Hellenistic model," we must first indicate the form that the conversion-to-the-self paradigm takes in this model. Foucault again begins with a discussion of differences between this stage and the preceding one, and he connects the main difference with the immanent nature of the *telos* that was just discussed. Conversion in Plato, represented in the metaphor of the Cave, is conversion-return, *epistrophē*, realized in hierarchical reality, the Platonic ontology of symbolic bi-une being, and is therefore characterized by the sharp opposition of two worlds: the one the soul has departed from and the one that constitutes the final goal of the soul's journey (the end result of the conversion, the *telos* of the practice of the self). According to Hadot's schema, which has become classical and which was accepted by Foucault as well, to this realization of conversion as *epistrophē* is opposed Christian conversion-*metanoia*, a "mind-change" that brings "complete revolution, radical renewal," etc. In the Hellenistic model, according to Foucault, conversion, here too, is a return to the self, but of a different kind than the Platonic *epistrophē*; it is a "conversion that is neither *epistrophē* nor *metanoia*" (244R; 209F). In contrast to the Platonic concept, this is a return that has become fully immanent: "Conversion . . . in the Hellenistic and Roman culture and practice of the self . . . is a return which occurs, so to speak, in the very immanence of the world. . . . Whereas the Platonic *epistrophē* consists in a movement that can carry us out of this world into the other world, what we are talking about now is a liberation in the very interior of this axis of immanence" (236R; 201-2F). It means that the ontological dimension of the paradigm of conversion, as well as of the entire practice of the self, is absent in the Hellenistic model. And this has direct consequences for subjectology. Return to the self now means not the discovery in oneself of one's "true self" which occupies a different position or even has a different status with respect to the surrounding empirical world, but nothing more than the acquisition of life-wisdom — of an indifferent attitude toward everything that does not depend on us. *In the absence of ontology, of an ontological structure of reality, there is no place for any "true self" that differs from one's "original self."* Therefore "the return to which Seneca, Plutarch, and Epictetus invite us is, in a certain sense, a turning in place: the only goal and destination is to establish oneself in oneself . . . and to remain there" (538R; 476F). Only one content remains in the conversion: a certain change in one's relation to one's self, which is unique, and always the "same." "The ultimate goal of conversion to the self is to establish certain forms of the relation to oneself. These forms are sometimes conceived in terms of the juridico-political model of self-sovereignty, of being one's own master. . . . They are also often represented in terms of the model of possessive enjoyment, of self-enjoyment, of taking one's pleasure with oneself" (538-39R; 476F). The distinction of this "ordinary," "everyday" *telos* from the *telos* of the Platonic-Neoplatonic practices of the self, as well as from the *telos* of spiritual practices, is obviously enormous.

Foucault also shows that, in the framework of the Hellenistic paradigm of conversion, an original cognitive paradigm arises: the paradigm of knowledge of the world treated as a constituent part of the actualization of conversion to the self. Contrary to the superficial view, the concentration on the self dictated by the Hellenistic practices does not exclude knowledge of the world, but rather includes it. This concentration on the self requires that all knowledge be subordinate to the "art of living," and this in turn means that knowledge in its totality must necessarily be transformed and ordered in conformity with this art: according to Foucault, "the Stoics insist that it is necessary . . . to direct one's gaze at oneself, while at the same time embracing the whole universe with it" (286-87R; 249F). It is worth paying attention to this last formula: it expresses the general principle of the construction of the totality of knowledge in an anthropological key and in an anthropological orientation (the orientation of "gazing at oneself"). Foucault analyzes in great detail how this principle is realized in Seneca: he describes the movement of the subject (or of his reason) who steps farther and farther away from himself, but "moving backward" as it were and not losing sight of himself — so that in the end this backward movement, embracing the whole sphere of the universe, "situates us at the peak of the world and thereby reveals before us the secrets of nature" (303R; 262F). And, according to Foucault, this Stoic paradigm is diametrically opposite to the Platonic one: "The Platonic movement of the soul . . . consists in turning away from this world and directing one's gaze at the other world . . . it leaves [this world] behind. . . . The Stoic movement of the soul is completely different . . ." (303R; 262F). "There is no transition to the other world. . . . This is not a movement that turns us away from this world . . . this is a movement that allows us — while not losing sight for a moment of the world, or of ourselves, or of ourselves in the world — to embrace the world wholly" (309R; 270F). Seneca himself and the other Stoics did not by any means see in their teachings a polar opposition to Platonism. Foucault corrects them, insistently playing his leitmotif everywhere: the purely immanent character of the Hellenistic practices of the self, of the culture of the self; the limiting of the Hellenistic worldview exclusively to "this world"; the adherence of Hellenistic man and reason to the idea that there is just the unique world, and all reality as such is limited to the world of our existence.

Of course, the main part of Foucault's reconstruction of the "Hellenistic model" of practices of the self comprises a concrete description of these practices. As already mentioned, the scope of Foucault's investigation includes the practices of three schools: Cynic, Stoic, and Epicurean. He does not, however, carry out a separate reconstruction of each of these three systems of practices; his analysis is conducted on a different structural level. Each of the practices is built out of specific units, out of different kinds of subpractices or exercises, intellectual, behavioral, physical, etc. Some of these subpractices are specific to a particular school, whereas others can be encountered in more than one school. It is these subpractices in their totality that constitute the direct object of Foucault's investigation. To designate these structural elements or units of the practice of the self he uses a term borrowed from Hadot: *Spiritual exercises*.

It is important to discuss this choice of term. In modern culture a single particular meaning has become firmly associated with it, where it is not just a term but has become a proper noun: in the sixteenth century it was the name that Ignatius of Loyola used to designate the system of the four-week Christian meditations he had created. In applying this term to the practices developed by the ancient schools of philosophy (primarily the Greek and Roman schools of late antiquity), Hadot asserted that, by doing so, he was only returning this term to its original sphere and original meaning; he also claimed that Loyola's use of this term could be traced back to its meaning in late antiquity. "[Loyola's] *Exercitia spiritualia* are only a Christian variant of the Greco-Roman tradition.... The concept and term *exercitium spirituale* were attested long before Ignatius of Loyola in early Latin Christianity, and they correspond to the *ascesis* of Greek Christianity. But, in turn, this *askesis,* which should be understood not as asceticism but precisely as the practice of spiritual exercises, already existed in the philosophical tradition of antiquity."[32] Foucault adopts this position without any discussion, speaking in passing about "the spiritual exercises which were commonly practiced in Christianity and which were derived from the spiritual exercises of antiquity, particularly of Stoicism" (320R; 281F). But it is worth noting that Hadot's comparativistic assertions about Christian borrowings are somewhat hyperbolic and not completely indisputable (a good example is his thesis about "the variant of the Greco-Roman tradition" created by a fiery young Catholic a thousand years after the end of this tradition!). However, at this time I will not analyze his assertions; nor will I discuss to what extent the concept of the "spiritual" in the sense employed by Hadot existed in the paganism of late antiquity (more than likely, it did not exist). In spite of all this, the main thing is indisputable, the existence of the cultural phenomenon itself: starting with antiquity, there is a dotted line of succession of goal-directed anthropological practices or exercises in which, according to Hadot, "not only the individual's thought but his entire psyche" participates and a "change of the vision of the world and transformation of the personality" are achieved. The name of the phenomenon, which has been given to it by Hadot and which has already gained wide acceptance, will be accepted as a fact also in the present analysis. A complete history of this phenomenon which goes all the way up to modernity has not yet been written; and Foucault, "receiving the baton" from Hadot, makes a number of remarks about its later stages which I will touch on below.

Foucault identifies a series of general characteristics that can be applied to the whole system of spiritual exercises. First and foremost among them is the concept of "equipment," *paraskeuē,* which conveys the purpose of all the exercises. As we have seen, in the examined practices of the self

---

32. Hadot, *Spiritual Exercises,* p. 23.

the "type of subjectness" and the nature and basic structures of the "self" do not change according to Foucault; in these practices "the path to the self" consists in a "turning in place." This turning signifies that only the *relation to the self* needs to be changed: it is necessary to establish adequate forms of this relation or, put more simply, it is necessary "to begin to relate to oneself in the right way." And this means that it is necessary to correctly see oneself-in-the-world, one's possibilities and tasks, and accordingly to develop and adopt correct and adequate principles of behavior, reactions, and strategies. As a result, these practices of the self are occupied, strictly speaking, not with the "self itself" as such, i.e., not with the transformation of the subject in his fundamental structure (Foucault insists on this, correctly seeing here a profound distinction from Christianity), but only with the subject-in-the-world, with the transformation of his strategies, reactions, and models of behavior. The practice (Foucault usually calls it "ascesis," or "philosophical ascesis," to distinguish it from the Christian kind) must equip the subject with models and reactions that will make him well prepared and adequate for all possible situations. It is this idea that is expressed by the concept of *paraskeuē*. "The chief and immediate objective of *askesis* . . . is the constitution of a *paraskeuē* (of a preparation, of an equipment) (354R; 312F) . . . *paraskeuē* is nothing else than . . . the ensemble of practices which are necessary and sufficient to make us stronger than anything that can happen in the entire course of our existence (349R; 307F). . . . It is an equipment, a preparation of the subject and the soul which arms them . . . in the due manner for all possible circumstances" (267R; 230F). The content of *paraskeuē*, the subject's "armor," is constituted by the Teacher's "true speeches," which must be assimilated by the Pupil in his own way, and must become not "received information" but the principles of the Pupil's action and behavior. The "true speeches" must be translated in the subject; they must be converted into some specific "active mode" and transformed into ever-present, ever-ready instructions of life-behavior (the Latin translation of *paraskeuē* is *instructio*). This is how Foucault describes the internal mechanism of the subject's "equipment": "*Paraskeuē* is a tool for the constant transformation of the true speeches which are rooted deeply in the subject into morally acceptable principles of behavior. . . . It is the element of the transformation of logos into ethos" (354R; 312F). As we see, here Foucault once again underscores the ethical nature of the Hellenistic model.

Further, a general view of the collection of spiritual exercises reveals different possibilities of organizing this collection. Foucault examines two principles of organization, which are opposite to a certain degree: an external principle, which refers to the spheres of human existence that are touched on, and an internal principle, which refers to the character of the exercises themselves. The external principle establishes a division into three parts: "The body, environment and household, love. *Dietetics, economics, erotics.* These are the three large domains in which the practice of the self is actualized in this [Hellenistic] epoch" (184R; 156F; italics added). According to Foucault, as early as Greece's classical epoch these three domains already represented the main spheres in the constitution of which the ancient art of living was realized: "... regimen, household management, the 'courting' of young boys ... [were the] three focuses ... [around which] the Greeks developed arts of living, of conducting themselves, and of 'using pleasures.'"[33] Here, "dietetics [was] understood as an art of the everyday relationship of the individual with his body ... economics as an art of a man's behavior as head of a family ... and ... erotics as an art of the reciprocal conduct of a man and a boy in a love relationship."[34] But at the same time in the classical epoch all three of these "focuses" did not enter into the sphere of the care of the self; on the contrary, "the care of the self ... in Socrates' speeches was distinctly separate from the care of the body (dietetics), from the care of the household (economics), and from the care of love (erotics)" (185R; 156F). Insisting as usual on a sharp distinction between the classical and Hellenistic epochs (and, by the way, greatly diverging with Hadot in this respect), Foucault sees the present distinction as consisting in the fact that the sphere of the care of the self and the practice of the self, cardinally expanding, absorbs the "three focuses" into itself: "Dietetics, economics, and erotics emerge now as the domains of the application of the practice of the self" (185R; 156F).

This external principle of the division of the practices and technologies of the self is convenient and is placed by Foucault at the basis of his study of sexual behavior (in particular, it determines the structure of the second volume of *The History of Sexuality*). However, in passing over to the hermeneutics of the subject and concentrating on the fate of the "self," Foucault finds the internal principle more relevant. He asserts, however, that with this division it has never been possible to fully organize all the spiritual exercises into a unified system — and this is due to their "philosophical" nature. "Philosophical life ... cannot be subordinated to any

---

33. Michel Foucault, *The Use of Pleasure: The History of Sexuality*, vol. 2 (Harmondsworth and New York: Viking, 1985), p. 249.

34. Foucault, *The Use of Pleasure*, p. 93.

rule, to any *regula*. It is subordinate to a *forma*, a form. . . . According to the ideas of the Romans or Greeks, subordination to rules, or just plain subordination, will never permit one to create a beautiful work. . . . And this is indisputably the reason why you will never encounter in the asceticism of the philosophers the kind of precise cataloguing of exercises . . . one finds among Christians. . . . Rather, one finds a much more confused ensemble [of exercises]" (459R; 406F).[35] He finds, nevertheless, that this "confused ensemble" has an internal structure: one can distinguish in it two "families" or classes of exercises of different types. He characterizes these types using two concepts from the vocabulary of practices of the self: *meletan* (to meditate) and *gymnazein* (to exercise or train). Exercises of the first type involve "a work of thought whose main function is to prepare the individual for that which he must soon do" (460R; 407F). Exercises of the second type, on the other hand, are real actions, "training in a real situation" in which, generally speaking, both the soul and the body participate.

Exercises of the second type have always been practiced; they are more than traditional for man since it was always characteristic of him to "verify what he is capable of." Foucault classifies such self-verifications into two types: exercises in abstention and (self-)tests. The first type includes training in fortitude, courage, endurance, tolerance of deprivations, but purely physical exercises (such as gymnastics and athletics) are now (in contrast to the classical epoch, to Plato) considered irrelevant for the practice of the self. However, even these time-hallowed exercises are rethought here: "In the culture of the self these exercises have a different meaning: to establish and to test the individual's independence in relation to the external world" (546R; 483F). A new image of the body is thus created: in place of the Athletic Body we have "the patient body, which has learned abstinence," "a body which has subordinated itself to the soul" — i.e., which incarnates, first and foremost, a definite "ethos of the body." (Self-)tests, which can take many forms, resemble the practices of abstinence, but in a number of respects they more deeply involve consciousness and thought. They include an element of self-knowledge; and inasmuch as typical tests involve exer-

---

35. This view, which reflects Foucault's idea of the "esthetics of existence," is not shared, however, by Hadot. He states that Stoic treatises "On exercise," with "a systematic codification" of spiritual exercises, existed but were lost; however, Philo of Alexandria has two lists of exercises, which Hadot reproduces: List 1: Seeking, deep study *(skepsis)*, reading, hearing, attention, self-possession *(enkrateia)*, indifference to indifferent things. List 2: Reading, meditation, healing of passions, recollection of that which is good, self-possession, fulfillment of duty (Hadot, *Spiritual Exercises*, p. 25).

cises aimed at suppressing one's emotions, they also include control of one's thoughts, "a special labor aimed at neutralizing one's thoughts, desires, and imagination." According to Foucault this labor approaches Christian dispassion, but without being as stark as the latter: "here we find ourselves midway." The concept of the "test" was gradually deepened: as in the evolution of the care of the self, here too there occurred a "generalization of the imperative," and the diversity of particular tests was complemented by interpreting the entirety of a man's life as a continuous test. In this interpretation the tests become, for all practical purposes, the same thing as spiritual exercises of the first type.

In the exercises of the first type the following general methodology (mentioned above) is implemented: certain occupations become spiritual exercises and are absorbed into the practice of the self by means of a special procedure that amounts to some turning or transmutation of them into a "practical" mode in which they are made into elements of a system for assuring the correct actions and behavior of the subject, his correct standing-in-the world. In the first place, this kind of turning should be performed for the generative concept of this entire group — for meditation itself, *meditatio*. This is by no means an ordinary "meditation on a certain thought"; it is an "exercise in the appropriation and assimilation of thought" as a result of which *thought must become effective and regulative for the acts and behavior of the subject.* Such assimilation implies, in particular, an "experiment in self-identification" — the immersion of oneself by thought into situations connected with the theme of the meditation. Here thought actively participates in the shaping of the subject, actualizing, according to Foucault, the "subjectivation of true speech"; and about meditation "turned" in this way one can say, together with Foucault, that this "is not a game of the subject with his own thought . . . but a game played by thought over the subject himself" (386R; 340F).

A similar "turning" is implemented in all exercises of this kind. As a result, "all technologies and all practices involving hearing, reading, writing, and speaking" become spiritual exercises. Foucault analyzes in detail three such exercises (or groups of exercises): hearing (and silence), reading/writing, and speaking (oral speech, *la parole*). In and of itself, hearing is the most passive modality, but in ascesis it is saturated with many active elements. In particular, it is transformed into "paraenetic hearing," which immediately translates that which is heard into the form of exhortation to the "self" and enters into the practice of the self as "the first step of the subjectivation of true speech." The main function of reading as a spiritual exercise was "to create an occasion for meditation," which had to realize the equipping of the self with "true judgments," and the latter, in their turn, had to be transformed into effective exhortations or instructions. Reading therefore relied on a strict selection of a few texts, usually fragments, and it was combined with writing: the primary purpose of reading was to collect judgments while that of writing was to transmute them into the necessary mode, and therefore, according to Seneca, "it was necessary to alternate reading and writing." Foucault was particularly interested in writing as a spiritual exercise; he devoted to it a special work, *The Writing of the Self* (1983). This exercise "reinforces and activates reading": it is a brief exposition of what has been read, a compilation of thematic summaries, of anthologies of excerpts, a record of useful conversations and lessons, and finally an epistolarium; and this whole corpus of what you have written constitutes "supports for the memory," *hypomnēmata*, playing, according to Foucault, an extremely important role in the culture and practice of the self of late antiquity. He calls it a "log book" of the "self" and sets for himself the task — certainly, an anthropologically necessary one! — of comparing such log books left by different epochs. Finally, speaking, the speech of the Pupil, of the subject, became an important element of the practice of the self only in Christianity, having taken the form of "practices of confession," of telling the whole truth about oneself; and, according to Foucault, "this feature was absolutely decisive in the history of subjectivity in the West" (392R; 346F). By contrast, in the Hellenistic practices "the one being led to the truth by the teacher's speech did not have to speak the truth about himself.... It was necessary that he be silent, and nothing else was needed"; all the elements of the Pupil's speech in these practices "had no spiritual significance" (392-93R; 347F). Only the Teacher's speech, which had to follow the principle of *parrhēsia*, had significance.

In addition, this type includes the group of exercises of philosophical analysis, so to speak. These exercises require that we observe our own representations and judge to what extent our judgments about them, as well as movements of our soul (passions and emotions) evoked by them, correspond to the truth. This is how Foucault defines their general content: "the stream of representations and the analytical work carried out to define and describe them," a work which demands that a representation is grasped in its spontaneity and that its objective content is identified. For the most part this work is accomplished by means of two exercises: clear seeing and description of the object in its structure ("eidetic meditation") and its naming, the saying of its name and the names of its constituent parts ("onomastic meditation"). Then, on the basis of the attained seeing of the object as such, it is necessary to attain its seeing in the general cosmic order: "to grasp the value of the object for the cosmos as well as its value for man as . . . a being placed by nature . . . in the interior of this cosmos" (324R; 285F). And in this cosmic seeing, according to Marcus Aurelius "the soul becomes great," approaching the Stoic ideal of imperturbability and indifference. Foucault separates and distances such an analysis of representations from the modern European philosophical methodology of the Cartesian type and instead indicates its closeness to the practices of the verification of representations in Christian ascesis.

Finally, the concluding group of exercises of the mature Stoic practice of the self is made up of complex and extensive exercises that sum up, as it were, the whole arsenal of the practice and its goals. Foucault considers three such exercises: examination of the soul; meditative prevention of evils (*praemeditatio malorum*); meditation on death. These exercises are widely known, and there is no need for us to discuss them at length. In his treatment of the examination of the soul Foucault contrasts it with the Christian practice: in their exercise of examination the Stoics employ not self-condemnation but rather a verification or test that measures one's distance from the goal, from the complete convergence in the "self" of the subject of truth and the subject of action. In analyzing the meditative prevention of evils, Foucault focuses on its temporal aspect, showing that what we have here is not at all an immersion in the future but rather an "abolition of the future," making possible a "reduction" of the whole sphere of evil to a "simple reality deprived of terrifying attributes." And Foucault concludes his lecture course with the meditation on death. The examination of the soul is given last in the transcript of the lectures, but the author's summary of the course ends with "the famous *meletē thanatou* . . . the exercise in dying." He calls it "the pinnacle of all exercises" and ends his text with a quotation from Seneca: "Death will show what I attained, and it is it I will believe."

I.2.3. The Christian, or "Religious," Model

The "Christian model" of the practice of the self, even though Foucault does not give a systematic reconstruction of it, is used by him as a constant example for comparison to the practices of antiquity in the most different aspects. The general contours and main accents of his treatment of the Christian practices therefore emerge before us clearly; he keeps repeating all the basic elements of his position, giving different variations of them and reinforcing them. I will begin the discussion of this treatment by surveying its real basis. What is the phenomenal basis on which Foucault makes his judgments about the Christian practices of the self?

To start with, he firmly adopts the thesis that the sphere of society and culture in which Christianity develops its practices of the self is chiefly the sphere of monasticism and ascesis. This sphere is cardinally different from the "philosophical ascesis" of pagan antiquity, even though it is connected to it by ties of close succession and abundant and numerous borrowings. According to Foucault, the Hellenistic-Roman thought of the first to second centuries clearly adopted the principles of the necessary self-transformation of the subject, focusing on the culture of the self, the care of the self, and the practices of the self. Therefore early Christianity, which posed the question of the transformation of the subject with unprecedented acuteness, could regard the Hellenistic-Roman thought as a preparation and anticipation of Christianity and could take much from it. "Christian spirituality, attaining its most rigorous form in asceticism and monasticism, starting in the third to fourth centuries, could present itself as the culmination of the ancient philosophy.... The life of ascesis, the monastic life, will be the true philosophy; the monastery will be the true school of philosophy" (202R; 172F). By moving succession and continuity to the foreground, Foucault downplays the impulse of the rupture with paganism, the role of which was indisputably dominant. I will return to this below. But while disputing Foucault's treatment of the relationship between Christian ascesis and "philosophical ascesis," I fully agree with his treatment of the significance of the former within Christianity. Foucault is perfectly correct when he notes the key role — unknown in the classical anthropological paradigm — of the ascetic-monastic tradition as the milieu, or laboratory so to speak, in which the Christian models of man, of I, of the self, as well as the Christian practices of the self, were created, verified, and guarded. "Beginning with the third to fourth centuries the Christian model [of the constitution of the subject and his relation to the truth] is formed. This model can more correctly be called an 'ascetic-monastic' model, rather than a Christian one in general" (281R; 244F).

Because of this general position, one would think that the main part of Foucault's "database" has to consist of early Christian monasticism, its ascetic practices. But the real situation is ambiguous. Foucault does in fact allot a great deal of attention to early Christian asceticism; he describes and analyzes it; but at the same time he relies on just one author of the ascetic tradition, John Cassian (c. 360-432). Only Cassian's texts are subjected to

PRACTICES OF THE SELF AND SPIRITUAL PRACTICES

Foucault's customary careful analytical reading.[36] Other Christian authors (but not ascetic ones) and their texts to which he refers are not analyzed but are only used fleetingly for purposes of illustration and to confirm certain theses of his position. (Not formulated on the basis of sources, this position therefore has a clearly a priori nature.) But there are even very few such authors who play for him the role of "servants for the occasion." More substantively than to the others he refers to Tertullian, more specifically to his small treatise *On Repentance*, which serves as Foucault's main source when he discusses the early Christian institution of public repentance. In addition, Foucault mentions Gregory of Nyssa's *On Virginity* repeatedly, in different texts; he regards this work as a major landmark on the historical path of the "care of the self," for Gregory's treatment of this concept, which relates it to the marriageless state, is used by Foucault to support his central thesis that self-renunciation is the core of the Christian relation to the self. All the other Christian authors are only mentioned in passing: Clement of Alexandria in a general comparison of ancient and Christian ethics and the culture of the self; Basil the Great as the compiler of the first monastic code; John Chrysostom and Augustine in an examination of individual spiritual exercises adopted by Christianity from the "philosophical ascesis" of the pagans; and Anthony the Great and Athanasius the Great in connection with the role of writing as the recording of phenomena and states of consciousness in ascetic experience. Finally, separate from this there is an imposing collection of sources on the Catholic practices of repentance and confession, primarily during the Counter-Reformation (sixteenth to eighteenth centuries), which serves as the basis of the two above-mentioned lectures of the 1974-75 Course. However, the "practices of confession *(l'aveu),*"[37] the theme that interested Foucault most in Christianity even in his later period, is not yet problematized through the prism of the practices of the self in the 1974-75 Course. As in all his investigations, Foucault used there his origi-

---

36. Needless to say, this assertion must remain provisional as long as Foucault's basic texts on the "Christian model" are inaccessible to us. But the assertion is confirmed by the "Brief Summary" of the 1980 Course and by the published fragment of *Confessions of the Flesh*.

37. When speaking about "practices of confession" Foucault uses the French term *l'aveu* (Russ. *priznanie*) and not the term *la confession* (Russ. *ispoved'*). Both terms are habitually translated into English as *confession*, but this way the different connotations of *l'aveu* (in terms of avowal or admission) and confession (in terms of the Christian sacrament) are lost. The present translation solves this problem by keeping the original French term *(l'aveu)* where appropriate. — Trans.

nal nonclassical approach, which conceptualizes the phenomena studied in terms of practices, in a "verb-type" active discourse; but the main practices for him then were the practices of power. The Course stated: "All these phenomena [phenomena of religious psychology such as visions, possession, etc.] . . . can be understood not in terms of science or ideology, not in terms of the history of teachings and theories . . . but in the framework of a historical study of the technologies of power."[38]

In sum, John Cassian is virtually the only representative source examined by Foucault substantively and in detail. Tertullian's *On Repentance* belongs to the period just before the author abandoned Christianity for Montanism, a sect of fanatical rigorists characterized by barbarically cruel practices of suppression of personality and mortification of the flesh (we will discuss this treatise below). Gregory of Nyssa's *On Virginity*, which praises the marriageless state, was written by him soon after his own marriage; with respect to the problems of family and marriage, the body and the life of the body, "the views of Gregory of Nyssa," as John Meyendorff writes, "are not characterized by clarity and bear the imprint of the influence of Origen, so beloved by Gregory."[39] Neither text possesses any particular significance or weight in Christian spirituality. There is no doubt that if one were to rely solely on the above-mentioned sources, it would be impossible to identify, reconstruct, and adequately understand the whole repertoire of the practices of the self created and cultivated in Christianity. Thus in what manner and to what degree is this repertoire reflected in Foucault?

Agreeing with the etymology (*askesis* = exercise, practice), Foucault accepts the view that the sphere of the practices of the self is, first and foremost, the sphere of ascesis: "philosophical ascesis" in the culture of late antiquity and monastic ascesis in the Christian world (the spheres of monasticism and ascesis in Christianity never fully coincided,[40] but Foucault does not go into such distinctions). He makes a whole series of comparisons of the two ascetic formations, and from this comparative discourse we can extract his general characterization of Christian asceticism. He formulates three "clear distinctions" between these formations, and in his formulations the following defining characteristics are attributed to Christian ascesis: (1) "the goal of this ascetic practice . . . is self-renunciation" (359R; 316F); (2) this type of

38. Michel Foucault, *Abnormal* (St. Petersburg: Nauka, 2004), p. 271. In French: *Les Anormaux* (Paris: Gallimard, 1999).

39. John Meyendorff, *Vvedenie v sviatootecheskoe bogoslovie* (Introduction to Patristic Theology) (New York: Religious Books for Russia, 1982), pp. 190-91 [in Russian].

40. See Horujy, *Phenomenology of Ascesis*, pp. 72-73.

asceticism institutes "a system of sacrifices, of obligatory renunciations of something in oneself and in one's life" (359R; 316F), and therefore (we read in another place) "the fundamental task of the Christian type of ascesis . . . consists in establishing what must be renounced, and in what sequence it must be renounced, in order to arrive at the final renunciation, the renunciation of the self" (526R; 465F); (3) "this ascetic practice of the self is based on the principle of the subordination of the individual to the law" (359R; 316F). It is considered that the main technology of the self in Christian ascesis (I will examine it later on) is the articulation and explication of one's own inner reality, or in Foucault's terminology, the "decoding" or "exegesis" of the self, which he interprets as the "objectification of oneself," and objectification means subordination to the law. This is how Foucault outlines this connection: "Early Christianity introduced several important modifications in the ancient asceticism: it intensified the form of the law but it also oriented the practices of the self in the direction of the hermeneutics of the self and the decoding of the self as a subject of desire. The law-desire coupling seems to be very characteristic for Christianity."[41] This last remark is valid here only if it is corrected: this coupling is of course characteristic — and even fundamental — for Judaism, the religion of the Law, whereas it is characteristic for Christianity as such only by virtue of succession, as a "reflected light."

In this series of distinctions the first characteristic is the chief and central one. The thesis that self-renunciation, the total renunciation of oneself (of one's individuality, self, identity, etc.), is the goal and meaning of Christian ascesis, of the whole Christian culture of the self — this thesis is the cornerstone of Foucault's position and the focus of his entire treatment of the phenomenon of Christianity; and he repeats it dozens of times, like an incantation, in all his texts. And all the other characteristics of the Christian culture of the self are connected in one way or another with this final and all-encompassing goal of it, are colored by and subordinate to this goal. Foucault's treatment of Christian ascesis is indisputably influenced by Nietzsche, who wrote most extensively about ascesis in the final section of *On the Genealogy of Morals*, titled "What Do Ascetic Ideals Mean?" We read there: "Will governed by the ascetic ideal . . . signifies a will to nothing, an aversion to life, a revolt against the most radical presuppositions of life."[42] Foucault adopted this thesis with exactitude, and in the context of

---

41. Foucault, "Le souci de la vérité," *DE II*, no. 350, p. 1491.

42. Friedrich Nietzsche, *K genealogii morali* (On the Genealogy of Morals) (Moscow: Mysl', 1990), p. 524 [Russian translation].

the conception of the practices of the self he developed it into the principle of self-renunciation. This central principle of the Christian practices of the self is sharply opposed by him to the practices of the self of late antiquity, whose goal was just the opposite: "to establish oneself as the desired goal" (284R; 247F), "to posit oneself — in the most explicit, strongest, and most stubborn manner — as the final goal of one's own existence" (359R; 316F).

According to Foucault, this opposition does not in any way contradict his view that Christian ascesis — in its entire structure and content, in its concrete exercises and working principles — descended from the Hellenistic ascesis, is the latter's successor and remains secondary in relation to it, and is thoroughly permeated with borrowings from it. In their genesis and development "spiritual practices in the Christian East involved an asceticism . . . which traced its origin back to the Stoics and Cynics" (455R; 403F). Moreover, even at the most general level the ancient "care of the self became something like a matrix of Christian asceticism" (22-23R; 12F). (Let me note that Foucault makes such a global assertion on the basis of just the title of one of the chapters of Gregory of Nyssa's *On Virginity*!) In addition, as we have already said, Foucault asserts, following Hadot, that the Christian model adopted from the Hellenistic one the structure of the practice of the self as an ensemble of "spiritual exercises" as well as the exercises themselves: "These exercises . . . were incorporated into Christianity and survived in it" (455R; 402F). In his analysis of concrete exercises, Foucault also concretizes his assertions about borrowing: from the Stoics Christianity inherits "the importance of exercises in abstinence . . . and exercises in self-knowledge, which Christian spirituality will cultivate . . . following the model of the old Stoic suspiciousness with regard to the self" (456R; 403F). In the same way, as Gros informs us, "in the lecture of 19 March 1980 Foucault develops the important thesis that Cassian renewed in Christianity the pagan philosophical technologies of control and testing in connection with the problem of preparing a hermit for his departure into the wilderness" (143R; 119F). And, inevitably, to borrowing in the sphere of the practices of the self there corresponds an analogous borrowing in the sphere of morality which is directly connected with these practices: "In the framework . . . of the Hellenistic model . . . a certain morality took shape which was inherited, adopted, and assimilated by Christianity and reworked by it into something which we call — illegitimately — 'Christian morality'" (285R; 247F).

In sum, for Foucault in essence the only new (autochthonous, authentically Christian) thing in Christianity is the supreme principle of self-renunciation (according to Foucault, the principle of self-decoding is not new, but has Stoic roots). Hadot agrees with Foucault's thesis, which insists that all the essential elements of the Christian practices of the self and moral principles were borrowed from the culture of (late) antiquity. But Peter Brown, another major (together with Hadot) authority on the culture of late antiquity, disagrees with it. Brown warns: "I have frequently observed that the sharp and dangerous flavor of many Christian notions . . . has been rendered tame and insipid through being explained away as no more than inert borrowings from a supposed pagan or Jewish 'background.'"[43]

Apart from these theses and assertions about the general characteristics of the Christian practices of the self, the accessible corpus of Foucault texts also contains a more detailed discussion of a series of themes. Foucault's attention was most drawn to the closely interconnected practices of confession *(l'aveu)* and practices of the examination *(l'examen)* of consciousness. According to the "Brief Summary" of the course *On the Government of the Living*, which is the point of departure and a major landmark for the whole program of the practices of the self, "the main part of the course is devoted to procedures for the examination of souls and confession *(l'aveu)* in early Christianity."[44] Under the term "practices of confession *(l'aveu)*" Foucault unites all the procedures both in the monastic and in the secular modes of Christian existence in which the Christian was required to present a "true speech about himself," a pure-hearted and detailed picture of his inner world: of his thoughts and movements of consciousness, of his desires and impulses, of his weaknesses and inclinations, etc. Foucault's heightened interest in these practices was inherited by his last project from the preceding period when his attention had been focused on the practices of power. Then his interest in the practices of confession was conditioned by the fact that he directly connected the phenomenon of confession with a situation of power and in no way restricted it to the religious sphere. In one interview, given after the lecture course *Abnormal*, we read: "[Question:] In one of your lectures you attempted to show that we live in a society of confession. . . . We have Christian confession, the confession of the communist, the confession of the writer, psychoanalytic confession, judicial confession. . . . Do all these confessions have the same structure?" "[Answer:] No . . . confession consists in the subject's speech about himself in a situation of power *(dans une situation de pouvoir)*, when over him there is someone's

---

43. Peter Brown, *The Body and Society: Men, Women, and Sexual Renunciation in Early Christianity* (New York: Columbia University Press, 1988), p. xvi.

44. Foucault, "Du gouvernement des vivants," *DE II*, no. 289, p. 945.

domination, he is limited in his actions, and with his confession he changes this situation.... That is the formal definition of confession, capable of embracing the types of confession you have mentioned."[45] As we can see, this "formal definition" is entirely located in the discourse of power. Now, in his new stage, Foucault regards this same type of practices as tending to belong to the practices of the self.

It is precisely the practices of confession that primarily express a key element of the Christian relation to the self: the *hermeneutic* nature of this relation. In the practices of confession a Christian actualizes the hermeneutics, decoding, exegesis of himself. Foucault underscores that this did not exist in the Hellenistic practices: although "suspiciousness" or mistrust toward oneself was present, as we have seen, among the Stoics, it was not a digging into one's inner reality but only a rectification of one's relation to oneself and by no means did it have cardinal self-transformation as its goal. According to Foucault, in the practice of the self the subject of antiquity, by assimilating "true speeches," becomes a subject who expresses the truth, so that "subjectivation of true speech" occurs, whereas in the practices of confession the Christian subject, by effecting verbal self-decoding, "objectifies himself in true speech." "Christianity ... requires of everyone ... that he apply himself to discovering that which is happening in himself, that he confess his errors and temptations and articulate his desires; he must then reveal these things either to God or to other members of the community, thus bearing witness ... against himself."[46]

It is clear that the empirical forms in which such practices are embodied consist, first of all, of the practices of repentance and confession, which are inseparably linked, as their necessary condition, with the practices of the examination of one's own consciousness, of everything that occurs or has its seat in it. These practices constitute the main topic of Foucault's studies in the sphere of Christian spirituality. Both the 1980 lecture course (according to its "Brief Summary") and the 1982 Vermont text discuss the same three basic themes: the early Christian institution of public repentance, or "exomologesis"; the monastic practice of the revelation of all one's thoughts to a spiritual director, or "exagoreusis" (which is part of the discipline of

---

45. Michel Foucault, "Les réponses du philosophe." *Dits et écrits* I, 1954-1975, no. 163 (Paris: Gallimard, 2001), p. 1677. This edition will subsequently be referred to as DE I, with indication of the text number and page number. An analogous but more extensive and profound definition of confession is given in the first volume of *The History of Sexuality* (London: Penguin Books, 1984), pp. 61-62.

46. Foucault, "Les techniques de soi," *DE II*, no. 363, p. 1624.

obedience and "cutting off of the will"); and the technology of the examination of consciousness.

According to Foucault, exomologesis and exagoreusis are "the two main forms of self-revelation" in early Christianity. The practice of exomologesis, an early form of repentance, was terminated by the church in the East by the end of the fourth century and by the church in the West only in the seventh century. Here is what it consisted of: a Christian who had sinned heavily requested that he be given the special status of a "penitent"; this status stayed with him for years and it implied punitive deprivations and limitations as well as a whole series of public rituals of humiliation and rules of self-abasing behavior. Analyzing it on the basis of Tertullian's treatise *On Repentance* (in the Course of 1980 he uses also *Apostolic Constitutions*[47]), Foucault indicates that exomologesis is "not verbal but symbolic, ritualistic, and theatrical"; it is "a theatricalized expression of the penitent's situation which demonstrates his status of a sinner." The focus of Foucault's attention here is the structure of consciousness and processes in the consciousness of the penitent. Of course, repentance "is a method for wiping away sins, of returning to an individual the purity he had acquired in baptism." It is also "a demonstration of change, of a rupture with oneself, with one's past and with the world." And the most essential thing is that "the goal of repentance is not to establish an identity but, on the contrary, to mark the renunciation of oneself, the rupture with oneself: *Ego non sum, ego*. This formula is at the core of *publicatio sui*. It represents the rupture of an individual with his previous identity. . . . Self-revelation is at the same time self-destruction."[48] It must be said that these last statements are not sufficiently supported by the sources Foucault refers to. These early sources are very poor when it comes to the analysis of consciousness: Tertullian is rhetorical and exalted in style but he is by no means ascetic in the sense of profound and unceasing introspection, the key characteristic of Christian ascesis. Moreover, the sphere of ascesis did not even exist in his era. As for the *Apostolic Constitutions,* they do not even touch on themes of consciousness. Therefore, Foucault's conclusions reflect in part the general characteristics of Christian repentance and in part his own interpretation of these characteristics, an interpretation guided by his reception of Christianity.

Exagoreusis, one of the specifically monastic technologies of the self,

---

47. A collection of ecclesiastical law dating from the fourth century.
48. Foucault, "Les techniques de soi," pp. 1626-27. For Tertullian, *publicatio sui* is synonymous with *exomologesis.*

is analyzed by Foucault on the basis of the texts of John Cassian, and he gives it the following definition: It is "an analytic and continuous verbalization of thoughts which the subject practices in the framework of a relation of absolute obedience to a master."[49] This technology, which was developed starting with the fourth century, in the early stages of Eastern monasticism (its transmitter in the West was John Cassian), is incorporated into the very heart of Eastern Christian ascesis, which will receive further shaping in the hesychast tradition. As Foucault's definition indicates correctly, there are two main aspects in the structure of *exagoreusis:* first, a verbal expression as complete as possible by the ascetic of all the phenomena and processes in his consciousness; secondly, the fact that this expression is in the form of a communication from the ascetic to his spiritual master. Both of these aspects are essential for Foucault's conception. The first of these aspects is closely linked with the practices of the examination of consciousness and self-monitoring of thoughts, and there is no need to examine it separately. As for the second aspect, the constant, maximally complete (to the point of trivial details) revelation of one's consciousness to the Other and the subsequent unconditional acceptance of his judgment over all the contents of one's consciousness presuppose, evidently, a very special relation to this Other. The *Elder-Novice* "anthropological dyad," a fundamental element of Eastern Christian ascesis, is precisely this kind of relation; it is a relation of absolute obedience. In describing it, Foucault contrasts its total, all-encompassing nature to the practices of antiquity: "The obedience that is required by monastic life . . . differs from the Greco-Roman model of the relation to the master in that . . . it involves all the aspects of monastic life. There is nothing in the life of the monk that can escape the sphere of this fundamental relation. . . . For all of his acts, even the act of death, the monk must ask permission of his director. . . . Obedience . . . presupposes total monitoring of behavior by the master. It is a sacrifice of self, a sacrifice of the subject's will. It is the new technology of the self."[50] The famous monastic technology of "the cutting off of the will" is described accurately here. There is no principle more alien to Foucault than obedience of any kind, and especially absolute obedience, but here he has enough academic impartiality and sharp-sightedness to see the personological fruitfulness of this principle. "Although the value of this relation depends on the qualifications of the master, there is no doubt that, in itself, the form of obedience, whatever its object may be, has a positive

---

49. Foucault, "Les techniques de soi," p. 1631.
50. Foucault, "Les techniques de soi," p. 1628.

value"⁵¹ — and it has value precisely for the construction of the subject: "The self *(le soi)* must constitute itself by obedience."⁵²

He is also sharp-sighted enough not to reduce the whole meaning of ascesis to obedience: "Obedience [is] far from being an autonomous final state."⁵³ In this connection there arise two questions: What is the true final state and what are the general contours of the whole ascetic path, of the ascetic practice of the self in its totality? But, alas, Foucault does not get far in answering these questions, and the direction he takes is not a completely correct one: the only thing he sees as the final state is *contemplation,* and he does not even clearly tell us what he understands it to mean. "The goal *(l'objectif visé)* is permanent contemplation of God.... It is a monk's obligation to incessantly turn his thoughts toward this point which is God and to assure himself that his heart is pure enough to see God."⁵⁴ But "to turn one's thoughts toward God" is a general religious principle that is not even necessarily Christian, and "purity of heart" as a precondition for seeing God is already mentioned in the Gospels (Matt. 5:8), and so we do not learn anything in particular here either about ascesis or about contemplation. Joined with obedience, this obscure "contemplation" constitutes for Foucault the binomial of the guiding principles of the whole of Eastern Christian ascesis: "The dominant principles of this new technology of the self, which is of Eastern origin and was developed in the monastery, are obedience and contemplation."⁵⁵ In reality, however, neither obedience nor contemplation can in any way be called a "dominant principle" of Eastern Christian (i.e., hesychast) ascesis, and there is no particular interconnection between them. Obedience is one of the instrumental, not one of the generative, principles of ascesis; as for contemplation, its role and the understanding of it differ in Western and Eastern Christianity and represent a separate complex problematic, which includes, in particular, the comparison with Neoplatonism (which Foucault avoids mentioning, even when it is necessary). Simplifying for the sake of brevity, I will only say that hesychast communion with God unfolds in the paradigm of personal communion, whereas contemplation tends to correspond to impersonal, speculative mystical experience; and although this term is frequently applied to hesychast experience, such an application is, strictly speaking, correct only when contemplation is con-

51. Foucault, "Du gouvernement des vivants," pp. 946-47.
52. Foucault, "Les techniques de soi," p. 1628.
53. Foucault, "Les techniques de soi," p. 1628.
54. Foucault, "Les techniques de soi," p. 1628.
55. Foucault, "Les techniques de soi," p. 1628.

ceived in a very special sense, i.e., when it is identified with communion and deification.

When he considers the theme of the examination of consciousness, our atheistic thinker is no longer wandering in the sphere of mystical experience as in a fog. Before us there once again emerges a confident Foucaultian analysis, identifying two "major forms" of examination, one of which resembles the practices of antiquity (every evening a review and analysis of the day that has passed), while the other is a new and more important form — "constant vigilance over oneself." This is how Foucault describes the famous ascetic — or rather hesychast — art of capturing and testing one's thoughts *(logismoi)*: "It is a question of grasping the movement of thought *(cogitatio = logismos)*, of examining it deeply enough so as to grasp its origin and decode where it comes from (from God, from oneself, or from the devil), and of sorting through all this (which Cassian describes using several metaphors, the most important of which is . . . that of the money changer checking his coins)."[56] This is a general schema whose every stage is, in turn, a special ascetic labor, a technology of the self, if you will. In the Vermont lecture Foucault unfolds this schema in greater detail, though not very successfully: "Sounding the depths of what is happening in himself, [the monk] tries to immobilize his consciousness, to eliminate the movements of spirit that turn him away from God. This implies that one examines every thought that presents itself to the consciousness in order to perceive the connection that exists between act and thought, between truth and reality, and in order to see if there is anything in this thought that is capable of rendering our spirit mobile, of provoking our desire, of turning our spirit away from God."[57] This statement is inaccurate. Foucault has a very good sense of the embryonic movements of thought in a monk's consciousness, movements that Western philosophy, as a rule, does not feel or know; but Foucault does not grasp what exactly it is that the ascetic consciousness does with these movements. What he says about "immobilizing the consciousness" is misleading. What is required is to eliminate not movement as such but a specific type of movement: what is required is to eliminate the mind's wanderings and scatteredness in order to make possible, intensify, and energize the mind's movement or ascent toward God. What is required is to change the type of dynamics of the consciousness, the type of organization and coordination of the energies of the consciousness, which in

---

56. Foucault, "Du gouvernement des vivants," p. 947.
57. Foucault, "Les techniques de soi," p. 1629.

synergic anthropology is called the type of man's energic image. But this entire subtle labor of spiritual practice is outside Foucault's field of vision.

In addition, at the sorting stage an important role is played by the above-mentioned verbalization of thoughts which is incorporated in *exagoreusis*. As we have seen, this verbalization must be a complete revelation of oneself to the master, a "confession" which presupposes a relation of absolute obedience. Deftly capturing these interconnections, Foucault identifies an ensemble that integrates different elements of ascetic experience and process: "Unconditional obedience, uninterrupted examination, and exhaustive confession form an ensemble in which each element implies the two others."[58] But what is the meaning of this ensemble? What is attained in it and what does it serve? These questions are crucially important, but it turns out that the answers to them are fully dictated by Foucault's a priori "ideological principles." First of all, all this is "an indispensable element in the government of some men by others as it was put into practice in monastic institutions."[59] And secondly this is required for "the constitution of such a relation to the self which tends to the destruction of the form of the self."[60] With this canonical thesis (which in this case, by the way, does not follow from his analysis) Foucault ends the "Brief Summary" of the 1980 Course.

I will now consider two or three other themes discussed by Foucault in the framework of the "Christian model." Given his special interest in sexuality, it is not surprising that in his examination of early Christian ascesis he gives a detailed analysis (although again on the basis of a single author, John Cassian) of "the struggle with the spirit of fornication." This is part of the struggle with passions, or "invisible warfare," the most ancient and well-known aspect of Christian ascesis; Foucault devotes to it a section (representing a systematic study) of "Confessions of the Flesh." Based on Cassian, the standard ascetic science is expounded here concisely and clearly: first, the place and significance of fornication (sexual sensuality, lust) among the other major passions, and its connections and relationships with them; secondly, the characteristics of this passion and its internal structure; thirdly, the struggle with this passion and the stages of its eradication. But, needless to say, in this exposition Foucault places the accents in his own way, makes profound observations, and presents to us, in conclusion, his own view of the topic. In the initial theme he variously underscores the important position and role

---

58. Foucault, "Du gouvernement des vivants," p. 948.
59. Foucault, "Du gouvernement des vivants," p. 948.
60. Foucault, "Du gouvernement des vivants," p. 948.

of fornication: "... fornication, while being one of the eight elements in the list of vices, finds itself in a particular position in relation to the others: at the head of the causal chain [connecting the vices], at the very origin of falls ... at one of the most difficult and decisive points of the ascetic combat."[61] There is a certain degree of hyperbole here: it cannot be maintained that the ascetic tradition (or even the source used, Cassian) ascribed such an extraordinary and conspicuous significance to fornication. Further, in discussing the structure and subtypes of fornication, Foucault sees the key feature of Cassian's position in the fact that he exclusively directs his attention at those manifestations of the passion that are within the limits of individuality. From this Foucault draws some important conclusions: "There is [in Cassian] no fornication per se. Absent from this microcosm of solitude are the two major elements around which revolves the sexual ethics not only of the ancient philosophers but also of a Christian like Clement of Alexandria ... : the coupling of two individuals *(synousia)* and the pleasures of the act *(aphrodisia)*. The elements put in play here are the movements of the body and of the soul, images, perceptions, memories, dream figures, the spontaneous course of thought, the consent of the will, waking and sleep. And two poles appear here which clearly do not coincide with the body and soul: the involuntary pole of physical movements or of perceptions ... and the pole of will itself which accepts or repulses, turns away or allows itself to be captured, takes its time, consents.... [Such are] the two forms of 'fornication' [to which Cassian wholly devotes his analysis]: *immunditia* [filth], which surprises a soul in waking or in sleep ... and *libido* [enjoyment of lustful feelings], which lurks in the depths of the soul ..."[62]

As this makes clear, even as regards a theme like the struggle with fornication, the first concern and tactic of ascesis is not the regulation of actions, the prohibition of certain acts, and so on, but a subtle monitoring of the structures of the consciousness and personality that makes possible the transformation of these structures. Cassian pictures this transformation as "six degrees of chastity," the first of which is "that the monk not be subject to excitation of fleshly lust in the waking state"; the second is "that the mind not occupy itself with lustful thoughts" and so on; the sixth and highest degree is "that even in sleep there be no tempting dreams about women."[63]

61. Foucault, "Le combat de la chasteté," *DE II*, no. 312, p. 1118.
62. Foucault, "Le combat de la chasteté," p. 1121. This quotation contains paraphrases from Cassian's *Conferences* (XII, 2).
63. John Cassian, *Pisaniya* (Works) (Moscow: Trinity Lavra of St. Sergius, 1993), p. 390 [Russian translation].

Foucault's analysis of this "ladder of perfection of chastity" (Cassian's formula) reinforces and deepens his view that these ascetical works represent a highly organized practice of the self. Examining all six degrees, Foucault concludes that an ascetic must "remain in relation to himself in a state of perpetual vigilance with regard to the smallest movements that can be produced in his body or soul.... This vigilance constitutes the implementation of a 'discrimination of thoughts' which is at the center of the technology of the self as it was developed in Evagrian spirituality.... This is an entire technology for the analysis and diagnosis of thought, of its origins, its qualities, its dangers, its powers of seduction, and all the obscure forces that can be hidden beneath the appearance that it presents."[64] It would be an inadmissible simplification to say that this technology merely produces an "interiorization of the catalogue of prohibitions," which replaces prohibitions of actions by prohibitions of thoughts and intentions. The essence of this technology is more profound and sutble: "In this ascesis of chastity one can recognize a process of 'subjectivation' which pushes far to the side a sexual ethics centered on the economy of acts. But ... this subjectivation is inseparable from a process of knowledge which makes the obligation to seek and speak the truth about oneself an indispensable and permanent condition; although it is a subjectivation, it implies an infinite self-objectification of oneself — infinite in the sense that ... it does not have an end in time, and in the sense that it is always necessary to carry through as far as possible the examination of the movements of thought no matter how small and innocent they might appear to be."[65] Foucault also connects this process of infinite objectification in the attainment of the "six degrees of chastity" with the above-mentioned paradigm of successive renunciations or rejections which he considers universal for Christian experience: "This separation [into degrees] indicates that progress in chastity is recognized according to the elimination of negative manifestations — different traces of impurity disappear successively.... One can interpret such progress as a task of dissociation.... The six degrees are six stages of a process in which the involvement of will [in the activity of body and soul] is terminated."[66]

It must be acknowledged that in this analysis of the ascetic "combat of chastity" Foucault manages on the whole to avoid the fixed ideas with which — appropriately and (more often) inappropriately — his entire reception of

---

64. Foucault, "Le combat de la chasteté," p. 1124, p. 1126.
65. Foucault, "Le combat de la chasteté," p. 1126.
66. Foucault, "Le combat de la chasteté," p. 1120, p. 1122.

Christianity is permeated: the ideas of power, suppression of personality, destruction of the self, and so on. Although conducted on a narrow and insufficient basis, his analysis penetrates sufficiently deeply into its object and thus convincingly accomplishes his strategy, which is to represent the "combat of chastity" as an ensemble of refined practices of the self. Peter Brown called this text of Foucault's a "brilliant sketch," and he might have given it this assessment not just because the sketch ends with a respectful reference to Brown himself.

By contrast, in the theme of pastorship Foucault's bent to reduce the Christian experience to relations of power is actualized with no constraints at all. In his work "The Subject and Power" (1982) this treatment of pastorship is presented with a clarity that does not require any commentary, and we can confine ourselves to quotations: "... Christianity proposed new relations of power and extended them over the whole ancient world.... We will call this technology of power, which was engendered in Christian institutions, pastoral power.... Christianity is the sole religion organized into a Church. And as a Church Christianity postulates the theory that certain individuals are called, by virtue of their religious quality, to make use of others ... in their capacity as pastors. This word ... signifies a quite particular form of power. 1) This is a form whose final objective is to assure the salvation of individuals in the other world. 2) Pastoral power is not just a form of power that rules its subjects; it must also be ready to sacrifice itself for the life and salvation of the flock. In this way it differs from sovereign power.... 3) This is a form of power which concerns itself not only with the community as a whole but also with each particular individual, over the course of his entire life. 4) ... [T]his form of power cannot be exercised if one does not know what goes on in the heads of people, if one does not explore their souls, if one does not force them to reveal their most intimate secrets. It implies a knowledge of consciousness and an ability to direct the latter. This form of power is oriented toward salvation (in contrast with political power). It is sacrificial ... and individualizing (in contrast with juridical power).... It is connected with production of the truth ... about the individual himself."[67] In other texts Foucault connects the mechanism of the exercise of pastoral power with practices of confession, trying to show that ideas of the sinfulness and guilt of every individual instilled by pastors served as the lever of power. Inculcation of the idea of sinfulness was served, in turn, by the *body-flesh* opposition put forward by Christianity in

---

67. Foucault, "Le sujet et le pouvoir," *DE II*, no. 306, p. 1048.

which the "flesh" embodied fallenness, susceptibility to lust and foulness. (Cf. for example Cassian's statement: "Not to feel the sting of lust ... would mean that one abiding in the body goes out of his flesh."[68])

Finally, let us note that in discussing the "Christian model" Foucault had inevitably to discuss the corresponding form of the paradigm of conversion (so important for him). On the whole, his treatment of this Christian form follows the guidelines indicated by Hadot's early works on the *metanoia-epistrophē* opposition. Cf. for example: "*Metanoia* ... is conversion with a total turning-over of the subject, who renounces himself and then is reborn" (242R; 207F). However, Foucault not only multiplies such formulations but also specially lays out the subjectological aspects of *metanoia*, which is of great importance for the hermeneutics of the Christian subject. Here is one more long quotation: "The word *metanoia* means two things: *metanoia* is repentance and it is also change, a radical transformation of one's way of thought.... Christian *metanoia* possesses the following distinguishing characteristics. First of all, Christian conversion implies a sudden change ... this change must be a unique and sudden event, both historical and metahistorical, which immediately reshapes and transforms the subject's mode of being. Secondly ... in this dramatic collapse of the subject taking place within history and also above history, you always have a transition: a transition from one type of being to another, from death to life, from mortality to immortality.... Thirdly, conversion can take place only if there occurs a rupture in the very interior of the subject. The self that converts itself is a self that renounces itself. To renounce oneself, to die to oneself, to be reborn in another self and in a new form which no longer has anything in common in its being, mode of being, habits, or *ethos* with the former self — this constitutes one of the fundamental elements of Christian conversion" (237-38R; 202-3F). Almost all of these features of the constitution of the Christian subject are already familiar to us. This conception of Christian conversion is one of the essential elements grounding Foucault's reception of Christianity, with its main thesis being the destruction of the subject.

* * *

For a long time, up to the stage of the practices of the self, Foucault's attitude toward Christianity was very much reminiscent of the stereotypes of European philosophical anticlericalism and, first and foremost, Marx-

---

68. John Cassian, *Works*, p. 76.

ism. The aim of the Marxist critique of religion is to expose the latter as one of the forms of class exploitation; Foucault substitutes "the technology of domination and repression" for "class exploitation," but he fully accepts the fundamental principle — the rejection of the proper irreducible sphere and proper foundations and roots of religion in the fundamental structure of the human being. Another frequent approach of secularist philosophy is the principle of "taking what is useful after rendering the sting harmless": secularist philosophy exploits concepts from the religious discourse after having emasculated their religious substance. Foucault applies this approach to the concepts of "spirituality" and "salvation," which he uses widely — in his own emasculated version. He also inherits the denunciatory and accusatory attitude; his words about Christianity and the church are often filled with the spirit and tone of *"écrasez l'infâme!"* But more important than all these "generic features" is the fact that Foucault had his own personal reason for adopting an accusatory stance toward Christianity, a reason that was the core of a profound and irreconcilable conflict with Christianity. This reason lay, of course, in the practices of confession. Foucault's treatment of these practices in his late period is almost dispassionate and academic and is far from reflecting all the aspects and peripeteias of the philosopher's relationship with this phenomenon. At the earlier stage he concerns himself with post-Renaissance practices of confession, which are "firmly imbedded in the practice of repentance," in the formal discipline of the church. Here he speaks of "the millennial yoke of confession" that took the form of "all-encompassing examination and trial"; he says that, "beginning with the Middle Ages, torture accompanies it like a shadow," so that, in sum, there appears before us one of the worst forms of the yoke, the suppression and crushing of personality. Foucault's attitude toward these practices, the constant concentrated attention he directed at them, bespeaks not only a professional position but also a profound personal and existential affect. One of Foucault's biographers, a member of his close entourage, writes: "Perhaps the very idea of 'confession' inspired Foucault with horror? Traces of this horror can be glimpsed in his last books, where he does his utmost to reject, censure, and expose the requirement that one express oneself, speak, confess."[69]

This is valuable testimony, but it is not quite correct in what concerns Foucault's life periods. Didier Eribon's biography came out in 1989, when

---

69. Didier Eribon, *Michel Foucault* (Moscow: Molodaya Gvardiya, 2008), p. 50 [Russian translation].

Foucault's late lectures had not yet been studied and there was no clear understanding of the "conceptual revolution" that had taken place in his last years. In reality, the "traces of horror" appear not in his last books but in the "penultimate" ones, whereas a major shift can be seen in the last ones. The scholar's gift and the depth of his vision compel him to see that the phenomena under consideration contain not only that which was so odious for him but also something entirely different, something valuable, and that in order to genuinely understand these phenomena one must turn to their sources and pose the questions differently. In the epoch of early Christianity and in the framework of the program of the practices of the self, the practices of confession disclose other aspects and possibilities, manifesting themselves as a kind of hermeneutics and self-hermeneutics of the subject, as a mechanism of the constitution of the subject, as a practice of the self capable of penetrating deep into the subject. In discovering such aspects and possibilities, Foucault is no longer inspired with horror, with the spirit of rejection and denunciation; that sort of spirit could hardly have produced "The Combat of Chastity." Not only does this result have a scientific importance, but it also represents an overcoming of oneself, an overcoming of stereotypes and complexes, a liberation from the image of the enemy.

But only to a certain degree. It is not by accident that Foucault always underscored the element of continuity in his work. The formula "conceptual revolution" should be used cautiously. It is indisputable that, at the new stage, the focus of attention was on the practices of the self. Foucault had liberated himself from the hypertrophy of the phenomenon of power, from *the oppressive power of power* over his consciousness; but as before, a cardinal significance is ascribed to the practices and relations of power, and, as Foucault finds, every type of the constitution of the subject is created "at the intersection of specific practices of power and practices of the self." Nor did he progress (or have time to progress?) particularly far toward a more profound, comprehensive, and dispassionate understanding of Christianity. Foucault is not just an investigator but also an apologist for the Hellenistic practices of the self,[70] and, in his analysis of the Christian practices, he is, in essence, not far from a position that, simplistically, can be expressed in the following words: *all the fundamental elements of the Christian practices of the self were borrowed from the practices of antiquity, and all the changes*

---

70. Cf., for example, Hadot's opinion: "His description of the practices of the self . . . is not just a historical study: he wishes to discreetly *(discrètement)* propose a model of life for contemporary man" (Hadot, *Spiritual Exercises*, p. 302).

that Christianity made in these practices were changes for the worse. And this refers, first and foremost, to the main change: to the fact that for Christianity the goal of the practice had become a rejection of the self, an absolute self-renunciation, realized through a successive (self-)disassembling of the subject. Having such a goal, these practices are of a defective kind that does not satisfy Foucault's definition of the goal of practices of the self: "to make of one's life a work which bears certain esthetic values and conforms to certain criteria of style."[71] A fundamental opposition thus arises: the practices of antiquity are practices of self-creation, whereas the Christian practices are practices of self-destruction. And only one conclusion is possible from this opposition: *Christianity is an anthropologically negative phenomenon.* That is the verdict that indisputably follows from Foucault's entire discourse about Christianity.

I.3. Outline of Foucault's "General Project" and Discussion of His New Conception of the Subject

There exists an inevitable logic: the new construction of the anthropological discourse, *the speech about man,* carried out in the program of the practices of the self implies a certain new construction — as well as a new interpretation — of the discourse of any science of man and thus of the whole sphere of humanities. These possibilities, which are implanted in the idea of the practice of the self, were clear to Foucault from the very beginning. In his Burlington lecture of 1982 he describes the full panorama of the humanitarian sphere as viewed from the angle of this idea. The new interpretation sees in the discourses concerning man certain "methods by means of which the man of our culture organizes knowledge about himself," and it attempts to disclose the anthropological roots of these methods, i.e., to trace them back to certain "technologies which people use to understand who they are." In this interpretation the sciences of man manifest themselves, in Foucault's terminology, as "games of truth" and as "imaginary sciences" (in the sense that their foundations do not lie in the things themselves but are derivative of anthropological practices, of man's relations with himself). The humanitarian sphere is therefore translated into the domain of anthropological technologies, and this domain possesses a natural structure. "[T]hese technologies are divided into four large groups each of which represents a ma-

---

71. Foucault, "Usage des plaisirs et techniques de soi," *DE II,* no. 338, p. 1364.

trix of practical reason: 1) technologies of production thanks to which we can produce, transform, and manipulate objects; 2) technologies of systems of signs which permit the use of signs, meanings, symbols, or signification; 3) techniques of power which determine the behavior of individuals, subordinate them to certain ends or to domination, and objectify *(objectivent)* the subject; 4) technologies of the self which permit individuals to effect, alone or with the assistance of others, a certain number of operations on their own bodies and souls, on their thoughts, their behavior, and their mode of being; and to transform themselves in order to attain a certain state of happiness, purity, wisdom, perfection, or immortality."[72]

Translating "human-dimensional reality" into the space of practice, Foucault poses in relation to it a global problem: "I wanted to describe both the specificity [of each type] of these technologies and their constant interaction."[73] Affirming, as usual, the conceptual unity of his work, he notes that this is a problem he considered not just in his last period but throughout his oeuvre; he says that this has been his "goal for more than twenty-five years," i.e., from his very first book. This clearly means that he regarded the conception of the practices of the self — and more broadly the treatment of anthropological reality as the ensemble of practices described above — as the foundation that he was furnishing to all his past work. That is also what Gros thinks: "Toward the end of his life the technologies of the self became for Foucault the conceptual crown of his entire oeuvre, something like final basic principle" (560R; 497F). However, the global problem is immediately narrowed; not all types of anthropological technologies are included in the orbit of the investigation: "The first two types of technologies ... apply to the study of the sciences and of linguistics. My attention was chiefly drawn to the two other types, to the technologies of domination and to the technologies of the self."[74] It is they that are responsible for the constitution of the subject, and so, as I have already pointed out more than once, "the general project" (as Foucault sometimes calls the total set of his designs) takes the form of a "history of the subject." This is what he wrote at the beginning of the 1980s in the article about himself for the *Dictionary of Philosophers*: "At the present time Michel Foucault, in the framework of his general project, has undertaken an investigation of the constitution of the subject as an object for itself: the study of the procedures by which the subject is led

---

72. Foucault, "Les techniques de soi," p. 1604.
73. Foucault, "Les techniques de soi," p. 1604.
74. Foucault, "Les techniques de soi," p. 1604.

to observe himself, to analyze himself, to decode himself, and to recognize himself as a domain of possible knowledge. In sum, the investigation concerns the history of 'subjectness,'[75] if this word is taken to mean the manner in which the subject obtains experience of himself in a game of truth in which he actualizes a relation to himself."[76]

This article explicates also something of no small importance: the methodological principles of the project. "If one takes the problem of the relations between subject and truth as the guiding thread of all these analyses, this implies a certain choice of method. And first of all a systematic skepticism with regard to all the anthropological universals, which does not mean that one should reject all of them right off the bat and for good, but ... everything that is proposed to our knowledge as having universal validity, with regard either to human nature or to the subject, must be proved and analyzed.... The first rule of the method ... is as follows: go around all the anthropological universals, inquiring into their historical constitution (not leaving out, of course, the universals of humanism which compel that the rights, privileges, and nature of a human being be regarded as an immediate and nontemporal truth of the subject). It is also necessary to invert *(retourner)* the philosophical ascent toward the subject when one asks to take account in the formation of his constitution of everything that can in general be the object of knowledge; on the contrary, one must redescend toward the study of concrete practices by which the subject is constituted.... From this we get a third principle of the method: to address 'practices' as a domain of analysis, to approach the study by a roundabout way, through 'that which one does.'"[77] Needless to say, there is no finished methodology in this collection of rules; it is easy to agree, however, that these or similar rules must be included in "prolegomena to any new anthropology."

It is just as easy to see how similar this project is to the ideas developed by Foucault at the beginning of his scientific path, in one of the earliest of his texts: the "Introduction" to the French translation of Ludwig Binswanger's book *Dream and Existence* (1954). In this "Introduction" not only is there nothing that anticipates Foucault's future war with the "anthro-

---

75. The French *subjectivité* can be translated into Russian either as *sub'ektivnost'* [subjectivity] or as *sub'ektnost'* [subjectness]. Following Foucault's meaning, I have used the second alternative, which successfully underscores that we are dealing not with a specific perspective introduced by the subject ("subjectivity") but with the fundamental structure of the subject as such, with a special "subject" mode of being, or "subjectness." — Trans.

76. Foucault, "Foucault," *DE II*, no. 345, p. 1452.

77. Foucault, "Foucault," pp. 1452-54.

pological illusion" and the "anthropological sleep," but on the contrary he begins it with a declaration, in which, without even having any prejudices against anthropology, one can suspect a kind of "anthropological illusion." This declaration of the young Foucault is a declaration of independence for anthropology, a declaration that puts forward the problem of the creation of an autonomous anthropological discourse that refuses to take its foundations from any existing sciences, but has them within itself. The following quote explicates this point very clearly: "These lines of my introduction are motivated by only one thing: the desire to present a form of analysis whose intention is not to be a philosophy and whose goal is not to be a psychology, a form of analysis which posits itself as fundamental in relation to all concrete, objective, and empirical knowledge, and, finally, whose principle and method are initially determined only by the absolutely privileged status of their object: of man or rather of man's being, *Menschsein*. In such a way, one can encompass *(circonscrire)* the whole bearing surface *(la surface portante)* of anthropology."[78] It is already clear to Foucault that one must progress toward the sought-for type of discourse through the element of concreteness and historicity, and here there arises for him the same image of the "roundabout way" as in the above-cited later text: "It seems to us that in contemporary anthropology Binswanger's approach corresponds to the royal road. It obliquely approaches in a roundabout way the problem of ontology and anthropology, going straight toward concrete existence, toward its development and historical contents."[79] The young Foucault "felt into" Binswanger's existential psychology the vision he had of the new anthropology; and he apparently believed then that this vision could be embodied in life by Binswanger's method, which he saw as an activity of thought armed with the arsenal of Husserl's and Heidegger's phenomenology and shuttling between *Menschsein* and *Dasein*, i.e., situated at the boundary of anthropology and ontology. However, dependence on even such an impressive arsenal did not satisfy Foucault and his thought sought other paths; and he soon changed his landmarks. The positions of *Les mots et les choses* can be considered, in the first approximation, diametrically opposite to those of the "Introduction," but even here he could have, I think, defended his thesis about the continuity and unity of his thought. Even before the last shift in his thought, in a 1975 interview given to Brazilian journalists, who couldn't refrain from asking him the sacramental question "What is man? Does anything like that

78. Foucault, "Introduction," *DE I*, no. 1, p. 94.
79. Foucault, "Introduction," p. 95.

exist?" — his answer was: "Of course he exists. What needs to be destroyed is that collection of valuations, specifications, and residue deposits which have been used to define human beings since the eighteenth century. My error was not that I said man didn't exist but that I imagined it was so easy to demolish *(demolir)* him."[80] In the practices of the self his anthropological intuition begins to acquire its own point of support, its own arsenal; and the later "general project" converges with the vision of his youth, with the initial project of 1954. Here his creative path acquires the distinct character of an *epistrophē*.

Such are the landmarks of Foucault's last project. He broke off work on it at an early stage, and at this point I will give a cursory outline of what he managed to accomplish. First of all, he clearly defined the general structure of the history of subjectness: this history (most fully developed in the 1982 Course) took the form of the succession of three major formations or models of practices of the self: the Platonic model in the paradigm of self-knowledge as self-recollection; the Hellenistic model in the paradigm of conversion to the self *(epistrophē)* as ethical self-cultivation; and the Christian model in the paradigm of "self-exegesis and self-renunciation" as religious self-cultivation. Foucault insisted on this triple schema, stubbornly affirming it as the basis of his conception: "It is this question of the constitution of the truth of the subject in its three major forms that I try to pose, perhaps with immoderate obstinacy" (280R; 243F). But of the three models only the second, the Hellenistic one, was fully reconstructed by Foucault, who gave it preference in many ways. He gave an in-depth analysis of the Platonic model only in those of its aspects that were connected with sexuality (we find this, first of all, in *The Use of Pleasure*), whereas he only cursorily examined its structures of subjectness and its other general-anthropological aspects.

Finally, for a number of reasons the Christian model has a special position. Although the main features of this model are delineated with great articulatedness, distinctly and decisively, nevertheless there is no overall clarity — first of all, because of *the problem of boundaries.* What is the relationship between Foucault's "Christian model of practices of the self" and Christianity as a historical (and anthropological) phenomenon? Foucault's model is formulated on the basis of early Christian ascesis of the fourth to fifth centuries as it is represented by John Cassian, with some isolated additions from Tertullian, Augustine, and Gregory of Nyssa. Besides this, Fou-

---

80. Foucault, "Les réponses du philosophe," *DE I*, no. 163, p. 1685.

cault analyzed the practices of pastorship and the practices of confession based on examples from other epochs and outside the domain of monasticism, chiefly using materials relating to the post-Tridentine Catholicism of the sixteenth to eighteenth centuries. Evidently, the conclusions of these analyses must also be considered as part of the "Christian model." However, the final picture is somewhat incoherent: there are too many unresolved questions. The model is based on the monastic practices of a certain narrow ancient period, and of a specific school to boot: as Foucault indicates in passing, it is based on the "Eastern monasticism" spread by Cassian. How widespread were these practices and what significance did they have beyond the limits of their period and their environment? What meaning will the "model" have if these practices are supplemented by ones that evolved not in early Christian monasteries but a millennium later in secular Western European society? The logic followed by Foucault when he constructs his model from such strangely chosen little islands consists evidently in the fact that all the phenomena he considers reinforce his conception of Christian subjectness as the destruction of the subject. It must be acknowledged, however, that this logic is not scientific but ideological in character. In the scientific respect the boundaries of the model are extremely fuzzy.

At the same time it is necessary to recognize that there are two horizons in Foucault's very unfinished project: wide and narrow. The narrower project, which for the most part is what Foucault has in mind in *The Hermeneutics of the Subject*, is limited to the frame of ancient history, which the philosopher indicated quite concretely: "I will try to trace the long genealogy of the relation of the subject to itself from *Alcibiades* to Augustine" (213R; 181F); or in another place: "[T]he extreme limits of our time frame are Socrates and Gregory of Nyssa" (536R; 474F). In this narrow project Foucault does not include his previous studies of the later Christian practices. The fuzziness that characterized the historical boundaries of the Christian model is therefore absent here, but the question remains of the meaning of this model beyond the limits of the Eastern monastic school. Foucault does not pose this question although it is an essential one: basing his analysis on both Cassian and Augustine, he does not (in any of his published texts) mention or take into account that these two fathers of the Church diverged in the cardinal question of grace and freedom of the will, a question that is far from irrelevant for the constitution of the subject. What is more, Cassian's position was officially rejected in the West (by the Second Council of Orange in 529) and declared to be a "semi-Pelagian heresy." The constitution of the subject according to Cassian (and in Orthodoxy) is clearly not the same as according to Augustine (and in Catholicism), which is why, in particular, the practices of confession in Eastern Christianity are very different from those in Western Christianity. And this means that Foucault's "Christian model" has reference to the Western subject but is in large part based on the data belonging to the Eastern subject, and fully ignores the profound difference between them. Thus it operates with an incorrect and poorly conceived database. As I will show in Section II, this model also does not take into account many other — and more fundamental — elements of the Christian practices of the self.

My remarks about the fuzziness apply not to the narrow but to the wide horizon of the project. Foucault's program in this horizon was the same: to reconstruct the history of subjectness as a process of the change of formations of practices of the self — no longer just up to Augustine, but up to the present. There is no doubt that the complete "general project" would have included this horizon; Foucault tells us this directly (see the quotation above); it is also evident in the constant orientation of his thought to modernity in virtue of which his historical models were, so to speak, automatically projected on modernity and were correlated with it (I have already presented Hadot's testimony about this); but it is most evident in his real ideas and studies, scattered throughout the body of his texts. Alas, these scattered elements do not allow us to reconstruct how Foucault imagined the history of the subject in its continuous development from Augustine (or Gregory of Nyssa) to our contemporaries. One can, however, make certain observations about his positions. The first observation is of a negative character: Foucault definitely does not think that the later history of the West saw the appearance of any new major formations of the practices of the self in addition to the three he identified in the Ancient world. In all his excursions into the domain of this history he only discovers surviving elements of these three formations, all of which have undergone more or less significant changes. Foucault exhibited a kind of anthropological pessimism: not having created new formations, man has thus not been able to make any new significant discoveries or contributions in the culture of the self over the entire course of the Christian epoch as well as the post-Christian epoch (which, judging by some of Foucault's remarks, begins, in his opinion, as early as Thomas Aquinas). If we look closely, we can see the roots of his pessimism. It is clear that the position that affirmed self-renunciation and self-destruction as the first principle of Christian subjectness already predetermined that Foucault could find in the Christian world only an absence of anthropological creativity. As for the post-Christian, secularized world, the conviction that it is also anthropologically barren is rooted in another important thesis of Foucault, which I have already mentioned above (see p. 18): the thesis that the principles of the care of the self and knowledge of the self have competed against each other in the history of the Western subject. The core of the thesis is that in this history Foucault saw not only a cultivation of the practices of the self but also another line of development: a *competing line*, where man rejected the practices of the self and preferred strategies where he did not occupy himself with himself but practiced objective knowledge.

In this line of development the practices of the self are abolished as it were because they are not necessary. "There is no need for the subject to transform himself. It suffices for the subject to be what he is in order to have, in knowledge, an access to the truth" (215R; 183F). Foucault stubbornly sought a pair of categories that would capture the essence of the opposition between the two lines: he contrasts them as lines of "spirituality" and "philosophy";[81] or of "cathartics," the discipline of self-purification, and "politics" (in the Neoplatonic sense of the terms); but the deficiencies of these two pairs are obvious, and he did not have time to find an adequate variant. All the same his texts make it possible to trace a "competing line" almost over the entire course of the history of the West, although all the stages and landmarks of its path are discussed, alas, only fleetingly. The common womb of the two lines is Plato, in whose thought both of them are already present, although they do not compete there but are identical to each other. The foundation of the separate line of "philosophy" is Aristotle, who was "the only philosopher of antiquity ... for whom the question of spirituality was not that important" (30R; 18-19F). The next landmark is theology of the scholastic type — "this theology ... which is based on Aristotle ... and which thanks to Saint Thomas, scholastics, etc. will take its place in the history of Western thought" (41R; 28F). It was the scholastic theologians who "drove a wedge" between "the possibility of access to the truth" and "spirituality," the principle of the necessity of the self-transformation of the subject: "Scholastics ... was already an attempt to lift the condition of spirituality to which all of ancient philosophy and all of Christian thought had adhered" (216R; 184F). Thus, according to Foucault there occurred in the history of Christianity a conflict between theology and spirituality, and this

---

81. In this case spirituality receives a particular (re)definition: it is "that search, that practice and experience, by means of which the subject works in himself transformations necessary for obtaining access to the truth" (p. 27R; p. 16F).

conflict was very prolonged: "In the course of twelve centuries ... from the end of the 5th century (St. Augustine, of course) up to the 17th century ... there was a war not between spirituality and science but between spirituality and theology."[82] And that decisive intensification of the "non-spiritual" line of development that Foucault calls the "Cartesian moment" became possible only on the basis of this prolonged stage that separated spirituality from "rationally grounding reflection." In this "moment," in Descartes' *cogito ergo sum*, "truth turned out to be accessible to the subject as such (215R; 184F).... Descartes said that for knowledge philosophy alone is sufficient" (41R; 27F); and as a result "the connection between 'spirituality' and philosophy was ruptured (I think, permanently)." The permanence of the rupture was reinforced by Kant: "Kant represents, if you will, one more coil of the spiral, which consists in saying that what we are incapable of knowing is precisely what constitutes the very structure of the knowing subject, which makes it so that we cannot know it" (215R; 183F). These two names signify, according to Foucault, the total triumph of the competing line of development: "Descartes and Kant abolish what can be called the condition of spirituality.... Kant and Descartes seem to me the two great moments here" (216R; 183F). And this triumph is the defining characteristic of the thought of Modernity: "The modern epoch in the history of truth began the moment when it was supposed that knowledge and only knowledge gives access to the truth" (30R; 19F).

In sketching this path of "non-spiritual philosophy," Foucault constantly makes qualifications; he sees the process as profoundly ambiguous: "In the 17th century there was posed the question of the relationship between the conditions of spirituality and the problem of choosing the path and method for arriving at the truth. There were many points of contact, many intermediate forms" (42R; 29F). And, what is even more important, the history of thought is dramatic: following the expulsion of "spirituality," there immediately ensue efforts to retrieve it. Foucault finds that the most significant phenomena of the philosophical thought of the nineteenth and twentieth centuries occur under the sign of these efforts. "Take the whole

---

82. We have before us a lecture, not a fully finished book, and Foucault did not correct himself here: needless to say, the separation of theology from spirituality on the basis of Aristotle, and more precisely on the basis of Aristotle in Aquinas's highly specific redaction, could in no way occur starting with "the end of the 5th century." Without Aristotle (the redacted version) this separation could not have acquired its foundation, and before the stage of scholastics, traces of the tendency to downplay "spirituality" could be seen only in a few theologians, such as (primarily) Abelard.

of the philosophy of the 19th century, or almost the whole of it, Hegel . . . Schelling, Schopenhauer, Nietzsche, Husserl of the *Krisis*, Heidegger again — and you'll see that here too . . . knowledge . . . is, as before, tied to the requirements of spirituality. In all these philosophies a certain structure of spirituality attempts to link knowledge, the act of knowledge . . . with a transformation in the very being of the subject. That's the only thing . . . the *Phenomenology of Spirit* is about. And the whole history of the philosophy of the 19th century can be conceived as a kind of pressure by which one tried to rethink the structures of spirituality into . . . philosophy" (43R; 29-30F). The fundamental vector of the philosophical process which in the nineteenth century was designated as the "reconstruction and second coming of the structures of spirituality" retains this direction also in the twentieth century. Husserl of the later period and Heidegger were already mentioned above as belonging to this orientation; and to a certain, albeit limited degree, Foucault puts Marxism and psychoanalysis in the same category. Fully aware that this sounds like a paradox, he writes: "I am by no means saying that these are forms of spirituality. What I am saying is that in these forms of knowledge you will find questions, problems, and demands which . . . if one looks at things in a historical perspective of at least one or two millennia, are very old . . . questions connected with *epimeleia heautou* and thus with spirituality" (44R; 30-31F). He also acknowledges that the problematic of spirituality together with all the approaches to it was veiled or masked here: "There were attempts to mask the conditions of spirituality proper to these forms of knowledge inside a certain number of social forms"; and because of this masking and substitution "one forgot the question of how the subject is related to the truth" (44R; 31F). As a result, even if spirituality was not totally absent here, it was very poorly represented. It is from these positions that Foucault assesses Lacan: he sees in him an attempt to overcome this defect, finding that "Lacan wanted . . . to re-center psychoanalysis precisely on the question of the relation between the subject and the truth," and as a result "on the ground of psychoanalysis he revived the ancient tradition . . . of *epimeleia heautou*, which had been the most general form of spirituality" (45R; 31F).

This is how Foucault's "wide project," sketched by him in a rough way, reaches modernity and Foucault's own position in it. Even though it is formed of large rough blocks, the schema has a certain coherence. After a millennium of the active cultivation of the care of the self which created the three major formations, dominance is acquired by the competing line of development, which cultivates "knowledge of the self," separates it from self-transformation, and finds its main support for this separation in Aristotle. In its initial, scholastic-theological stage this line pushes aside the "condition of spirituality," the principle of self-transformation; at the next, philosophical stage, with the arrival of Descartes and Cartesianism, it directly rejects this condition and principle: this, according to Foucault, is philosophy of the "classic type," the philosophy of "Descartes, Leibniz, and so on." Then there begins the reverse process of "the reconstruction and second coming of the structures of spirituality" to which most major thinkers and currents of the nineteenth and twentieth centuries attach themselves in one way or another. It is not surprising that Foucault regards his own thought as converging with this "second coming." In certain texts of his last years he characterizes this line of "the reconstruction of the structures of spirituality" in other terms: the question of the conditions of our access to the truth is evidently also the ontological question about us ourselves as to what we are in our present-day reality; he calls this question, thus posed, "the problem of the ontology of the present *(ontologie du present)*, the ontology of ourselves." And he quite definitely puts himself in the philosophical current that occupies itself with this problem: "... a critical thought which will take the form of an ontology of ourselves, of an ontology of actuality, is a form of philosophy which, from Hegel to the Frankfurt school while passing through Nietzsche and Max Weber, founded a form of reflection in which I have tried to work."[83] In another similar passage the line of succession is given in greater detail: "... at the end of the 18th century ... philosophical activity conceived a new pole, and this pole is characterized by the permanent and perpetually renewed question, 'What are we today?' ... Kant, Fichte, Hegel, Nietzsche, Max Weber, Husserl, Heidegger, and the Frankfurt school attempted to answer this question. Inscribing myself in this tradition, I am trying to give ... answers to this question through the history of thought or, more precisely, through an analysis of the history of relations between our reflections and our practices ..."[84]

The historico-philosophical aspects do not exhaust all the dimensions of Foucault's project as an anthropological project; the schema described above does not give an answer to the most essential questions relating to the destiny of the practices of the self. As Foucault himself underscores, the line of "the reconstruction of the structures of spirituality" does not yet accomplish any actual reconstruction: all it does is to "pose, at least, implic-

---

83. Foucault, "Qu'est-ce que les Lumières?" *DE II*, no. 351, pp. 1506-57.
84. Foucault, "La technologie politique des individus," *DE II*, no. 364, p. 1633.

itly ... the question of spirituality," and it does not return to the principle of care of the self but only "manifests, without saying it, the care of the care of the self" (43R; 30F). In other words, Foucault is referring to certain fairly embryonic features and to tendencies that are not too pronounced, and it is not surprising that the boundary between the line of "the reconstruction of spirituality" and the "classical philosophy" which (according to Foucault) has rejected "spirituality" turned out in many cases to be arguable and uncertain. If the criterion of "spirituality" is the recognition that for the attainment of truth in knowledge it is necessary for the subject to change, then some changes or other are presupposed by any cognitive paradigm, and I am not convinced by Foucault's attempt to show that the changes which are part of the classical paradigm of Cartesian type are only inessential changes that "do not concern the subject in his being ... the subject as such" (31R; 19-20F). In sum, the position in Foucault's schema of all the three main figures of classical European metaphysics — Descartes, Kant, and Hegel — is highly arguable. Descartes is assigned the role of the pillar that supports the whole line that rejects the "condition of spirituality," but Hadot more than once argued eloquently against such a treatment; he writes in one of his books that "we find spiritual exercises also in Descartes (at least in the *Meditations*),"[85] while in another book he points out, directly addressing Foucault, that "in choosing the word 'Meditations' as the title of one of his works, Descartes knew perfectly well that in the tradition of ancient and Christian spirituality this word meant exercise of the soul. ... Each of the six 'Meditations' represents a spiritual exercise, i.e., the work of the thinking I on itself, a work which is necessary for passage to the next stage. ... In Descartes we still have the ancient conception of philosophy — particularly in the letters to Princess Elizabeth, which ... are an example of spiritual directorship."[86] Analogously, Hadot asserts that Kant too belongs to this conception of philosophy. But, in the case of Kant, Foucault himself repeatedly refined and made richer the treatment given in the 1982 Course, which was obviously too simplified and one-dimensional. In the 1983 Course, a fragment of which was published in the form of an article, it is stated that this treatment corresponds to the philosophy given in the "Critiques" (in another place Foucault limits it to the *Critique of Pure Reason* alone), whereas

---

85. Hadot, *Spiritual Exercises*, p. 356.
86. Pierre Hadot, *Chto takoe antichnaya filosofiya?* (Moscow: Izdatelstvo Gumanitarnoi Literatury, 1999), p. 278, p. 280 [Russian translation]. In French: *Qu'est-ce que la philosophie antique?* (Paris: Gallimard, 1995).

in other texts of Kant one can see the beginnings of "another method of critical questioning" which leads to an "ontology of the actual" and grounds one more "great critical tradition" — the very same one that produces "the reconstruction of the structures of spirituality" and in which Foucault places himself.[87] Finally, the inclusion of Hegel in the line of "the reconstruction of spirituality" is not less arguable than the inclusion of Kant in the line of its negation. There are grounds to believe that this inclusion was based for Foucault on the reception of Hegel that was developed by Kojève, adopted by a student of Kojève's lectures and Foucault's teacher Hyppolite, and became greatly influential in French thought. But this Kojèvian reception of Hegel's thought (first and foremost, of the *Phenomenology of Spirit*, which is the work that Foucault too always primarily has in mind) as anthropology represents nothing more than a highly specific point of view, which has not been accepted, for example, in the German tradition and which does not in any way supplant certain quite different receptions (such as that of Kierkegaard), which repeatedly pointed out the profoundly anti-anthropological character of Hegel's thought. From these other points of view, which are not less well-grounded than the Kojèvian one, one would have to put Hegel in the other line of development.

Thus, looking at it closely, Foucault's historico-philosophical schema turns out to be somewhat shaky: the cursory jottings left by the philosopher do not convincingly achieve the goal he set for himself, i.e., the goal of describing two distinctly different and antagonistic lines of development in the European philosophical process, one of them accepting (explicitly or implicitly) the principles of the care of the self and the practice of the self, the other rejecting them in favor of "pure knowledge." But at the same time his general picture of the process can be accepted: it is indisputable that in its post-ancient history, European thought went away from the above-mentioned principles and engendered — first in theology, then in philosophy — conceptions and entire major formations that were sharply divergent with them, and that the nineteenth to twentieth centuries saw the emergence and the growth and intensification of tendencies to regenerate them in new, still vague forms (which in the present day are often conveyed by the formula "anthropological turn"). There is also no disputing that all the philosophies discussed by Foucault are still very far from directly restoring the rights of the care of the self and from including into their orbit the practices of the self, in the same way as these practices were included in

---

87. Foucault, "Qu'est-ce que les Lumières?" *DE II*, no. 351, p. 1506.

the philosophy of the Stoics or in ascetic theology. On the other hand, Foucault, like Hadot, is firmly convinced that the principles of the care of the self and the practice of the self (spiritual exercises) have retained their universal significance and are valuable and necessary for the present day, for modern man. Therefore if one looks through the texts of both philosophers one can find, in the form of sporadic and not-too-noticeable digressions, the theme of *the modern fate of practices of the self*. This theme is developed in the style of free and nonrigorous reflections and has a personal sound, bears an imprint of personal inclinations. As I have said more than once, Foucault's inclinations are associated with the Hellenistic practices of the self. Accordingly, it is around these practices that all his meditations on the practices of the self in the modern world revolve.

Here is an important passage from the 1982 Course. In examining the fundamental paradigm of the Hellenistic practice of the self, "the return to the self," Foucault finds that this paradigm and this practice implied a specific "ethics of the self and esthetics of the self," after which he presents a meditation on the paths of the culture of the self and the ethics of the self in subsequent epochs. It might seem strange, but this meditation is clearly guided by a naïve mythologem of the Golden Age, which compels one to see the whole history of thought as a series of constantly renewed "attempts to re-create a certain esthetics and ethics of the self," where the ethics of the self of late antiquity is essentially regarded as the unique standard. "In the 16th century you will find a real ethics of the self as well as a real esthetics of the self . . . which openly refer back to the works of Greek and Latin authors . . . the history of the thought of the 19th century can be reassessed from roughly the same positions . . . the thought of the 19th century can be perceived . . . as a series of . . . attempts to re-create an ethics . . . of the self" (277R; 240-41F). What is most essential, we have here not a simple scientific pinpointing of a repeating phenomenon or typologically related phenomena, but precisely the orientation toward a unique and permanent standard, toward a prototype of the ethics of the self that remains the unique and necessary standard even in the present day: "The creation . . . of such an ethics in the present day . . . is . . . an urgent, essential, and politically indispensable task" (278R; 241F). However, here too Foucault demonstrates anthropological pessimism already noted above: although the standard is clear and "the essential and urgent task" has been defined, Foucault has doubts about whether it can be fulfilled: "It is not excluded that the collapse of the renewed attempts . . . a series of more or less fruitless efforts . . . will engender a reasonable doubt whether such an ethics is possible in the present day" (278R; 241F).

It is clear, however, that the thought of a historian and methodologist of the caliber of Foucault will never leave such "naïve" patterns of thought unnoticed and not reflected upon. The above-described meditation is only a moment in Foucault's complex development of the large theme: "Hellenistic and the modern culture of the self." Here is another moment that corrects the first: in the lengthy in-depth interview "On the Genealogy of Ethics" (April 1983) we find the following subheading: "Why the ancient world was not a Golden Age but what we can nonetheless take away from it."[88] And the true picture gradually emerges. The Hellenistic practices of the self really do serve for Foucault as a unique and permanent standard; and the solutions and strategies he reflects on and proposes as orientation for the modern world presuppose the regeneration of the general type and essential features of these practices. But at the same time he does not conceive the development of these modern strategies as a retro-utopia, as a mere return to the past and a reproduction of what has already been. Such a return is excluded by two major interconnected factors: first, the radical otherness and newness of the texture of modern anthropological and social reality compared with antiquity, and secondly the presence of fundamental structural imperfections, of defects in the practices of late antiquity (as we will see, Foucault did not idealize them).

Let me begin with the second factor. Throughout his entire life Foucault exhibited an acute and almost anarchic rejection of all external compulsions, rules, and codes imposed on the existence of the individual (on the other hand, he always affirmed the fruitfulness and necessity of the most severe internal discipline, the most severe "self-mastery"). This allergic sensitivity to compulsion enabled him to detect in the structure of his beloved practices a profound contradiction they had not been able to overcome. At their earlier stages these practices clearly lacked anthropological universality: they were cultivated in a narrow stratum of society (narrow even within the stratum of free citizens), were recognized to be a sort of privilege of the elite, and "were grounded on the idea of social superiority and contempt for others" (F. Gros). But when the consciousness of anthropological universality is gradually formed in late antiquity, it is expressed, almost primarily, in a normative and compulsory morality, which Foucault rejected categorically. "The search for a form of morality which would be

---

88. Foucault, "À propos de la généalogie de l'éthique: un aperçu du travail en cours," *DE II*, no. 326, p. 1206. NB: this version of the interview is different from that cited above, *DE II*, no. 344.

acceptable for everybody — in the sense that everybody would have to submit to it — seems catastrophic to me."[89] He identifies this aporia in his last interview, "The Return of Morality," and it is precisely in connection with this aporia that he pronounces his famous saying: "It seems to me that the whole of Antiquity was a 'profound error.'"[90]

As for the first factor, it will become clear when we describe the anthropological (and ethical) strategy that Foucault proposed. This strategy is usually designated by the formula "esthetics of existence," which is not less widely popular than the "care of the self" and the "practice of the self," but which, in contrast to them, continues to the present day to evoke many disagreements and divergent understandings. I will therefore try to be precise. For Foucault this term has two spheres of usage: first of all it designates the defining characteristic of the ancient practices of the self and ancient morality, and only afterwards is it brought into modernity. This characteristic consists in the fact that, in the practices of the self, one strives "to make of one's life a work which bears certain esthetic values and conforms to certain criteria of style" (I have already quoted part of this formula.[91]) Shaped by this striving, these practices bear in themselves that which is called the esthetics of existence. Observance of this esthetics was understood as a moral task, so that, according to Foucault, the ancient practices of the self included "morality as the esthetics of existence." (Cf. ". . . classical antiquity's moral reflection concerning the pleasures was . . . directed toward a stylization of attitudes and an aesthetics of existence."[92]) Foucault supposes that this characteristic could be present due to the fact that in ancient society religion was not an institution that constituted the consciousness of the individual and regulated the whole ensemble of his practices; it was rather an insignificant factor in the fundamental structure of individual existence. In today's post-Christian society it has again become an insignificant factor, and Foucault therefore puts forward the hypothesis that the esthetics of existence could be transported and incorporated into modern anthropological and ethical strategies. Here is how he describes this important transition from antiquity to modernity: ". . . in Greek morality, people concerned themselves more with their moral behavior, ethics, and relations to

---

89. Foucault, "Le retour de la morale," *DE II*, no. 354, p. 1525. The recognition that late Stoicism did not avoid elements of compulsory morality gradually redirected Foucault's attention to the Cynics, who could not be suspected of such morality.

90. Foucault, "Le retour de la morale," p. 1517.

91. See Foucault, "Usage des plaisirs et techniques de soi," *DE II*, no. 338, p. 1364.

92. Foucault, *The Use of Pleasure*, p. 92.

themselves than with religious problems. . . . What preoccupied them most, their great theme, was to create a sort of morality which was an esthetics of existence. And so I ask myself if our problem today isn't in a certain way the same, since for the most part we do not believe that a morality can be founded on religion and we do not want a legal system which would intervene in our personal and intimate moral life."[93]

Thus, the idea of the esthetics of existence, derived from the ancient practices of the self, becomes a modern anthropological and ethical project, a project directed at creating "a sort of morality which would be an esthetics of existence." How far did this project move forward? Not too far, of course; nevertheless, in Foucault's last texts we do find some of its theoretical principles and certain ideas and proposals relating to its realization. As for the theoretical principles, in the same important conversation at Berkeley in April 1983, Foucault puts forward some radical propositions that perhaps are not fully dependent on the ancient ethics and practice of the self and that mark the boundary between the anthropological and ethical situation of the Greco-Roman epoch and that of the West in modern times. "We are not faced with a choice between our world and the Greek world. But since we can observe that certain of the great principles of our morality were connected at a given moment with an esthetics of existence, I think that historical analysis can be useful. . . . I believe that we must free ourselves of the idea of a necessary analytic connection between morality and other structures, social, economic, or political ones. [Question:] *But what kind of morality can we develop today . . .* ? [Answer:] What astonishes me is that, in our society, art is connected only with objects, not with individuals or with life. . . . Could the life of every individual not be a work of art? Why are a painting or a house objects of art, but our life is not?"[94]

As is easy to see, this dense text contains two decisively asserted theses. The first is a total distancing, an "untying" of ethics from all dimensions and constraints of social life; it is a declaration that the sphere of ethics ("of our individual morality, our everyday life," as Foucault refines it) has no relation to all the aspects and structures of the social apparatus. The second thesis is of opposite kind; it is the just-as-total "tying" of morality to esthetics. This tying is asserted defiantly and audaciously: the interviewer's

---

93. Foucault, "À propos de la généalogie de l'éthique: un aperçu du travail en cours," no. 344, p. 1430.

94. Foucault, "À propos de la généalogie de l'éthique: un aperçu du travail en cours," no. 344, pp. 1435-36.

question about the character of morality gets an answer where morality is not mentioned at all, but there is a counter-question which says by the rule of contraries that the life of every individual must be a work of art. And if this does represent an answer to the question, What must morality or ethics be?[95] — this answer announces that *there is no ethics except esthetics*. One can consider this second thesis the quintessence and concise formula of the project of the "esthetics of existence." The thesis is immediately reinforced by a historical argument: Foucault asserts that the main aspiration of the man of antiquity, his "great theme," was "to create a type of morality which would be an esthetics." Finally, in this Berkeley conversation Foucault expresses a certain demarcation which is important for him and which can be regarded as the third principle of the project. This demarcation relates to two connected modern phenomena: to psychoanalysis and to that which Foucault calls the "modern cult of the self" and his interlocutors call "concentration on oneself, which many consider the central problem of our society." It would appear that Foucault and his project are also speaking of concentration on the self, but here there are two different kinds of concentration. I have already explained the distinction between them; this is the old distinction relating to the nature of conversion to the self in the Hellenistic and Christian models of the practices of the self. In the first case Foucault asserts that there takes place only a "return to the self" — the establishment of a correct relation to the "self," while the latter remains essentially unchangeable; in the second case there takes place a "decoding of the self," which penetrates down to the final depths of the latter and leads to a radical change of the self. In the present day this distinction appears once again: the esthetics of existence affirms the Greek relation to the "self," but according to Foucault both psychoanalysis and the "modern cult of the self," although they are wholly post-Christian phenomena, inherit in relation to the "self" the Christian paradigm: "In that which can be called the modern cult of the self the goal *(l'enjeu)* is to discover one's true 'I' while separating it from that which can obscure or alienate it and decoding its truth using psychological knowledge or psychoanalytical work. . . . [N]ot only do I not identify the ancient culture of the self with . . . the modern cult of the self, but I even think they are diametrically opposed. What has occurred is exactly a reversal of the classical culture of the self. This took place

---

95. Generally speaking, Foucault distinguishes between ethics and morality (see, for example, Deleuze, *Negotiations*, p. 133, p. 151), but in this context this distinction does not play a role.

in Christianity when the idea of the self which it was necessary to renounce ... replaced the idea of the self that was to be constructed and created as a work of art."[96] Thus, this ancient relation to the "self," intensified to the point where the internal content and fundamental structure of the "self" are completely closed off and untouchable — this relation is the third principle of the esthetics of existence.

In addition to theoretical principles, Foucault also put forward certain practical considerations. Chiefly, these are ideas and proposals expounded in his interview in the gay press and relating to the restructuring of the culture of the self in the gay community. First of all, Foucault implants into the consciousness of this community the above-mentioned "third principle." He tries to convince it to renounce attempts at "self-decoding" in favor of self-realization in one's homosexual givenness — accepted as a givenness not subject to doubt: "... it is necessary to defy ... the tendency to reduce the question of homosexuality to the problem of 'Who am I? What is the secret of my desire?' Perhaps it would be better to ask oneself: 'What relations can be established, invented, multiplied, and modulated through homosexuality?' ... [H]omosexuality is not a form of desire but something desirable. We must be passionate about becoming homosexuals ... ."[97] Further, it is proposed that the esthetics of existence be inculcated by abandoning the primitive model of "homosexuality ... in the form of immediate pleasure, in the form of two young men meeting in the street, seducing each other with a look, putting their hands on each other's butts, and going their separate ways after a quarter of an hour." It is proposed instead that we "advance in homosexual ascesis ... [and] create a homosexual mode of life" based on "friendship, that is, the sum of all things by which we can give pleasure to one another."[98] "Mode of life" and "friendship" are put forward as key concepts in Foucault's reflections and proposals relating to the homosexual culture of the self. "To be gay ... is to seek to define and actualize a certain mode of life."[99] There must be a multitude of modes of life in which a multitude of types of inter-human relations must be cultivated: "There exists the relation of marriage and family relations, but there are many other relations that should have the possibility of existing. ... We must achieve the recognition of relations of temporary cohabitation, of adoption. ... [Question:]

---

96. Foucault, "À propos de la généalogie de l'éthique: un aperçu du travail en cours," no. 344, p. 1443.
97. Foucault, "De l'amitié comme mode de vie," *DE II*, no. 293, p. 982.
98. Foucault, "De l'amitié comme mode de vie," pp. 983-84.
99. Foucault, "De l'amitié comme mode de vie," p. 984.

Of children? [Answer:] Or why not of one adult by another? Why shouldn't I be able to adopt a friend who is ten years younger than I? Or even ten years older? . . . [W]e must try to imagine and to create a new legal field, in which all possible types of relations would be permitted to exist and not be hindered or abolished by . . . [existing] institutions."[100] To construct and organize these relations, to invent new ones — that is the creative element of the esthetics of existence. As for friendship, its particular flowering and the abundance of its models, organically including homosexual relationships, are what, according to Foucault, characterized the Hellenistic world, and this is exactly the reason why he declares: "I have a passion for the Hellenistic and Roman world before Christianization."[101] As a whole, the esthetics of existence appears here in the form of the "culture of gays" — a specific type of culture that "has meaning only on the basis of a particular sexual experience" and "makes out of pleasure the point of crystallization of a new culture."

In the makeup of the conception of the practices of the self the "esthetics of existence" is the least scientific part; it is a project and proposal addressed to a wide society and, properly speaking, to every contemporary. It therefore had its separate fate, being debated in wide circles, engendering more than a few misunderstandings, and producing disagreements even among those who were most knowledgeable. "The esthetics of existence is the source of so many errors," says Gros, and all his discussion of this project is developed as a refutation of these errors. In the foreground he places the indisputable fact that the ancient esthetics of existence (and consequently its modern version as well) is developed in the framework of the care of the self and the practices of the self, and he reminds us that, according to Foucault, to the Hellenistic practices of the self corresponds the strictest and most self-limiting morality. With this he refutes the main errors, which regard the esthetics of existence as "a concession to the temptation of Narcissism" and reach the point of "declarations that Foucault's morality consists in a summons to a periodic overstepping of boundaries, in the cult of inexhaustible marginality." It cannot be denied, Gros remarks correctly, that "Foucault is neither Baudelaire nor Bataille. In his last texts there is neither the dandyism of exclusivity nor the poeticization of transgression."[102] But

---

100. Foucault, "Le triomphe social du plaisir sexuel: une conversation avec Michel Foucault," *DE II*, no. 313, pp. 1128-29.

101. Foucault, "Le triomphe social du plaisir sexuel: une conversation avec Michel Foucault," p. 1129.

102. Gros, "On the lecture course," p. 578R; p. 511F.

these arguments are scarcely a sufficient response to the entire existing and possible criticism of the esthetics of existence. Hadot directs an insistent critique at both domains of the latter: he does not agree with the interpretation of the ancient practices and ancient philosophizing as an esthetics of existence; nor does he accept this esthetics as a modern model of existence. "Unlike M. Foucault, I would not speak of the 'esthetics of existence' with reference either to antiquity or to the philosopher's task in general. . . . In Platonism, as well as in Epicureanism and Stoicism . . . it is a question not of the construction of 'I' as a work of art but on the contrary of the transcending of 'I' or . . . of an exercise in which 'I' is situated in the totality and is experienced as part of this totality."[103] As for modernity, "the ethical project of the 'esthetics of existence' proposed by Foucault to modern man . . . seems to me too narrow, insufficiently taking into account the cosmic dimension inherent in wisdom and, in sum, constituting rather a new variant of dandyism."[104] Gros, contra Hadot, wishes to separate Foucault's project from dandyism; but Foucault himself included dandyism among the forms or phenomena of the esthetics of existence. (Cf. for example: "The idea that the main work of art to which care should be devoted . . . is the self *[soi-meme]*, one's own life, one's own existence . . . is encountered in the Renaissance . . . as well as in 19th century dandyism.")[105]

But it's hardly fruitful to argue about small things like the assessment of dandyism; the three principles of the esthetics of existence I have identified make it possible to assess Foucault's anthropological project as a whole. The second principle, "there is no other ethics except esthetics" (and I insist that this is the correct reading of DE II, 1436!), is very eloquent: if Foucault is really affirming esthetics in the capacity of ethics, then all his indisputable (here Gros is right) differences from dandyism, from Baudelaire and Bataille, are nonetheless differences within the same kind of morality, a morality which asserts that the moral is that which is beautiful. The differences are in esthetic tastes only. That which is beautiful for Foucault is not that which is beautiful for Baudelaire and Bataille (or the "Californian cult of the self," which he distances himself from), but that which is beautiful for the Stoics; and his morality is not an esthetics of transgression but an esthetics of severe moderation, of training oneself for trials, for death. But the first

---

103. Hadot, *Spiritual Exercises*, p. 286, p. 288.
104. Hadot, *Spiritual Exercises*, p. 321.
105. Foucault, "À propos de la généalogie de l'éthique: un aperçu du travail en cours," no. 344, p. 1443.

and the second and the third thing are all variants of the esthetics of existence. Further, Gros defends the esthetics of existence against the accusation that it corresponds to "thought stuck at the esthetic stage" (in the sense of Kierkegaard's schema where consciousness evolves from the esthetic to the ethical stage and then to the religious stage). Here he is absolutely right: the esthetics of existence indisputably corresponds not to the esthetic but to the ethical consciousness, but to a peculiar one, to one which, as a result of a well-thought-out intellectual and existential choice, identifies itself with the esthetic consciousness and affirms esthetics as the supreme and obligatory type of ethics (so that Kierkegaard's schema is inverted).

But the third principle leads even further into the heart of the project. If the esthetics of existence presupposes an esthetic and creative cultivation of some basis of the "self" taken as the given, not subject to any decoding, and not allowing anybody to "dig" in it, this position implies, first of all, a definite type of subjectness, or a model of the human being's constitution. Rejected here is self-verification and self-constitution through self-opening or self-unlocking in the encounter with the Other or, in Hadot's variation, in the integration of the self into the all-embracing cosmic whole. Accordingly, what is constituted here is not an open but a closed type of subjectness, an entirely immanent subjectness. This accords with the second principle, meaning that the esthetics of existence is an ethical consciousness that has chosen for itself the position of an esthetic consciousness: for, according to Kierkegaard, closedness (manifested precisely in the preservation of the closed, inaccessible core of the self) is the defining predicate of the esthetic consciousness: "the aestheticist ... remains constantly concealed ... however frequently and however much he gives himself up to the world, he never does it totally, there always remains something that he keeps back."[106] One should clarify that the type of subjectness in Foucault's project can best be characterized as a kind of "secondary closedness," for what we have here is not an under-ripeness, not "the first stage on the path" to self-unlocking, but a conscious choice and negative response to the possibility of unlocking. Further, one must ask: What for Foucault is included in this untouchable and unopenable basis of the "self"? The later texts give us sufficient material for an answer. The first and most important element is sexual orientation; the homosexuality of homosexuals should be carefully protected and untouchable for the institutions of society, for psychoanalysis, for ev-

---

106. Søren Kierkegaard, *Either-Or*, vol. 2, trans. Walter Lowrie (Princeton: Princeton University Press, 1974), p. 327.

erything. This initial core finds a natural expansion in the type of relations of the subject to his own body, methods for obtaining pleasure, methods for starting and sustaining relations, the types of relations that the subject establishes and into which he enters, and so on and so forth. There thus arise in society groups of subjects possessing similar fundamental structures. One such group is the gay community and we can assume that for Foucault the gay community was indeed the prototype of such a group. According to Foucault, any such group or community must be recognized as a freely existing and developing subculture, and all the institutions of society must assure for such subcultures a maximum of possibilities for their favorable existence. The gays are not the only subculture examined by Foucault; there is a long conversation almost wholly devoted to S/M, i.e., sadomasochism. He recognizes the same nature in it; it is another variant of the esthetics of existence, one that takes a valuable and interesting direction and indicates new pathways: "S/M is a real subculture. It is a process of invention . . . it is an erotization of strategic relations . . . a real creation of new possibilities of pleasure. . . . S/M practices show us that we can produce pleasure by means of very strange objects, using certain odd parts of our bodies, in the most unusual situations. . . . One very important thing is the possibility of using our body as a possible source of a multitude of pleasures. . . . We can create new pleasures."[107] Thus, the S/M subculture indicates graphically what should be the new mainstream of the development of man and culture; and it becomes clear what other subcultures need to be created: "Drugs as a source of pleasure must become an element of our culture. We must study drugs . . . and produce *good* drugs, capable of producing very intense pleasure. . . . Today drugs are part of our culture. Just as there is good and bad music, there are good and bad drugs. And just as we cannot say that we are 'against music,' so we cannot say that we are 'against drugs.' [Reply:] *So, the goal is the experiencing of pleasure and its possibilities.* [Answer:] Yes."[108]

And so the gay subculture, the S/M and drug subcultures . . . Any method of obtaining pleasure can be the core of the esthetics of existence if only its adepts do not formalize and institutionalize it, but instead, by creatively developing it, make of it a work of art. It is astonishing that nowhere does Foucault say or even hint that there is some limit, some line that

---

107. Foucault, "Michel Foucault, une interview: sexe, pouvoir et la politique de l'identité," *DE II*, no. 358, pp. 1561-62, p. 1557.

108. Foucault, "Michel Foucault, une interview: sexe, pouvoir et la politique de l'identité," p. 1557.

shouldn't be crossed; on the contrary, he insists that the acknowledgment of such a line would go against the whole logic of his thought. By this logic, there can be subcultures of cannibals, of serial murderers, etc.; indeed, are they so different from the sadists so joyfully approved by Foucault? Yes, a man must subject himself to severe self-limitations, but they must be strictly immanent and self-motivated. Many interesting things can be derived from this, for example, that the Sixth Commandment, "thou shalt not kill," is as unacceptable (independent of its content) as all the others, *because it is a commandment.*

In short, for Foucault the esthetics of existence implies a restructuring of the human community into subcultures of a specific type. Their specificity lies in the fact that they are constituted by their method of obtaining physical pleasure, i.e., by a purely biological and anthropological principle. They have another distinguishing characteristic, underscored by Foucault: they are fundamentally extra-institutional, are not constrained by any social institutions, do not establish their own, and are not subordinate to existing ones. With such characteristics one can see in them human genera or tribes sui generis; and one can say that in its social dimension the project of the esthetics of existence manifests itself as a kind of *neotribalism*. This "neotribalist model" has an important socio-anthropological characteristic. Its basic concept, that of subculture, clearly corresponds to the sociological concept of "minority," which here becomes also an anthropological concept: the anthropology that corresponds to the project of the esthetics of existence is an "anthropology of minorities." What role, however, is given here to the "majority"? In the social plane the "majority" is clearly the source of institutions and practices of the suppression of minorities, a direct analogue to the "oppressor class" in Marxist ideology, and the minorities must use all means at their disposal to achieve liberation and independence. But in the anthropological plane of Foucault's thinking the "majority" is excluded and vanishes from the anthropological discourse, which is virtually wholly constructed as speech about minorities.[109]

This elevation of minorities and cancellation of majority is the ex-

---

109. As is easy to notice, the principle of the elimination from anthropology of the discourse of the "majority" is decisively put forward in all of Foucault's sketches of the anthropological project for modernity. As for his capital studies of ancient anthropological formations, this principle is present there too, but it is not an emphasized and conspicuous tendency. Thus, in the second volume of the *History of Sexuality* his analysis of the sphere of erotics is entirely focused on pederasty, the "courting of boys," while heterosexual erotics is mentioned only in passing as an irrelevant triviality that does not merit analysis.

pression of a logic congruent with the entire history of Foucault's thought and life. For most of his life Foucault was a staunch defender of the rights of minorities. But the best defense is an offense, and in his anthropological project the defense of persecuted minorities takes the form of a position whereby all "majority" ceases to exist. We see in the last interviews and articles that the whole Territory of Man in this project appears as a territory of subcultures/minorities. But there is at least one significant consequence of such a successful defense: together with the "majority" there also disappears from the discourse a specific group or class of concepts, namely, all generalizing and universalizing anthropological predicates such as "Man as such," "universal man" (an important concept for Kierkegaard), the universality of man, universal humanity, etc. Even outside Foucault's project, in the contemporary anthropological discourse such concepts have long had an archaic sound and are used less and less frequently, but they have not been explicitly rejected. Such a rejection would be a landmark in the history of Man, in the development of his relations with himself: it would be a significant step toward dehumanization and the Post-human.[110]

Thus, in its radical discarding of the basic features of all traditional models of man and family and society, the neotribalist model has a perceptible post-human color. Of course, in connection with Foucault's "Nietzscheanism," which he was fond of proclaiming, it can be recalled here that Nietzsche too spoke of the "breeds" of people and that this theme of his engendered the crudely vulgarized Nazi construct of the "Aryan race." If not for the racist factor, which is profoundly alien to Foucault, this "Aryan race" could also be regarded as one of the subcultures of his model. (Deleuze traces Foucault's last project directly back to Nietzsche and characterizes it as "vitalism based on esthetics.")

It is also worth noting that, like virtually all conceptualizations of contemporary post-human trends, the neotribalist model does not yet have a serious philosophical grounding. At its basis is the principle of pleasure: every tribe/subculture defines itself by the method it uses to obtain pleasure. But Foucault analyzes this principle neither in the philosophical nor even in the historical plane, even though he usually does so with all of his concepts. He even ignores all discussions of pleasure in the Hellenistic epoch, first of all, in the polemics between Stoics and Epicureans, when, for example, the Stoic Hierocles (according to Aulus Gellius, "a worthy and authoritative

---

110. By "post-human," I mean a new and radically different being, like, e.g., a cyborg or mutant, that has committed a rupture with human constitution.

man") asserted that "to regard pleasure as a goal is worthy only of whores." Foucault does not pose the problem of how pleasure relates to desire and to other basic attitudes and structures of consciousness. In sum, in the discourse of the esthetics of existence pleasure appears as a pure ideal, similar, for example, to "communism" in the socialist utopia, and this enhances noticeably the general utopian quality of the project with which the thinker consoled himself in his last years.

This description of the esthetics of existence makes it clear that this project cannot be understood adequately without taking into account Foucault's own homosexuality. Of course, the influence of this factor is not limited to this particular project. Foucault — as he tells us himself in numerous texts — was not just a homosexual, but a militant and passionate homosexual, a man for whom sexuality played an enormous role in the course of his life and in the structure of his personality. It is specifically about his last period, with which we are concerned, that his biographer writes: "Foucault hungers to live wholly by his homosexuality."[111] "Wholly" must include all his work, especially because he was a philosopher who so intensely revived the understanding of philosophy as a mode of living. And in the light of all this, should one regard as so absurd the approach (criticized by this same biographer) that seeks "to interpret every line in Foucault based on his homosexuality, as many American university scholars attempt to do"?[112]

I do not have the intention to follow this approach, but for our discussion of the conception of the practices of the self it is nonetheless important to show that the influence of Foucault's homosexuality can — or must — be taken into account in two key points of this conception, one being the constitutive element of the Christian model of the practices of the self, the other being the constitutive element of the Hellenistic model. As regards the Christian model, the key lies in the "practices of confession." I have already indicated that it is precisely in confession that one can identify the core of Foucault's conflict with Christianity. It has now become clear that a homosexual factor is present (and, maybe, even dominant) in Foucault's attitude toward it: the homosexual consciousness categorically rejects and, according to Foucault, must reject all self-decoding and especially compulsory and public (in one degree or another) self-decoding, or "confession." And thus Foucault's homosexuality contributes greatly to the formation of the bundle of his strong emotions connected with the "practices of confes-

---

111. Eribon, *Foucault*, p. 349.
112. Eribon, *Foucault*, p. 48.

sion" — protest, fury, horror (above we presented a striking quotation from Eribon about them). In the plane of ideas, not emotions, the sharp hostility to the phenomenon of confession leads to a totally erroneous understanding of its nature and function, at least in the sphere of the ascetic tradition. This will be shown in Section II.

As for the Hellenistic practices, there was a significant emotional component in relation to them too, but of an opposite kind. We have already cited Foucault's words about his "true passion" for the Hellenistic and Roman world, and it is precisely in an interview on themes of homosexuality that he says this. The homosexual consciousness naturally tended to see in Christianity an image of the enemy, whereas in the Greek world it tended to see an image of a friendly world in which the models of existence and personhood admitted, as Foucault says, "all possible types of relations," including homosexual ones. What should these models be? In speaking about the problems of gays, Foucault repeats again and again: "The homosexual movement today is in greater need of an art of living than of science or scientific knowledge . . . about what sexuality is."[113] A homosexual should not study his homosexuality, but should live with it, take it as the foundation of his life and build an "art of living" on its basis. But the necessary premise of such attitudes is a specific type of subjectness, of the constitution of the subject. Evidently, it is that very same type of subjectness which is affirmed by the "third principle" of the esthetics of existence: the purely immanent constitution of the self which, in its fundamental structure, should be accessible neither to decoding nor to transformation. This type of subjectness is necessary for the homosexual consciousness as Foucault sees it, and he finds it precisely in the Hellenistic practices, as I have shown above. His treatment of these practices is characterized by an insistent assertion of their total immanence, their pathos of immanence, if you like; and we have now revealed the homosexual correlation, so to speak, of this important feature. But there's more. The validity of Foucault's assertion relating to the purely immanent type of subjectness in the Hellenistic practices turns out to be arguable: it is disputed by Hadot. In examining Foucault's analysis, he discovers in it "a great many inaccuracies," and he finds the chief inaccuracy precisely in the treatment of the constitution of the subject. Foucault's treatment is developed on the basis of the *Letters to Lucilius*. But Hadot finds that, in Seneca, this constitution is by no means immanent. "The Stoic finds

---

113. Foucault, "Michel Foucault, une interview: sexe, pouvoir et la politique de l'identité," *DE II*, no. 358, p. 1554.

his joy not in his 'I' but, as Seneca says, 'in the best part of himself.' ... 'The best part of oneself' is, in the final analysis, the transcendental self. Seneca finds his joy not in 'Seneca' but in transcending Seneca."[114] As we see from this, according to Hadot (whose authority in the study of late antiquity is indisputable) the constitution of the subject for the Stoics is not immanent, but transcendent. Why does Foucault ascribe immanence to it? As we can now conclude, his insistent tendency toward purely immanent subjectness is very probably rooted in his homosexuality.

*   *   *

Finally, in concluding this overview of the conception of the practices of the self, I feel it necessary to consider one strange detail of it. I mean here the relationship between Foucault's conception and Kierkegaard's philosophy, a relationship that can be resumed briefly as follows: the dependence of Foucault's new conception of the subject on Kierkegaard's ideas, concepts, and modes of thinking is, without any exaggeration, enormous, but Kierkegaard's name cannot be found anywhere in Foucault's later texts. The analysis of this unspecified relation between the two philosophers is not only a problem for the corporation of Foucault scholars, it is also relevant in this context since the relation in question also reflects certain characteristic aspects of Foucault's attitude toward Christian thought in general. I will not offer an in-depth analysis at this point and will instead confine myself to indicating the most important overlaps, in order to assess the general character of the interaction of Foucault's thought with the philosophy of the "Danish Socrates." My treatment of his philosophy will be based on my analysis of it in the book *Diogenes' Lantern*.

The main overlap is a borrowing of the main thing: there are sufficient grounds to say that the very conception of the practice of the self is taken from Kierkegaard. I have already noted this in the initial description of the practice of the self. I have also pointed out the connection of Foucault's conception with the ancient tradition of philosophizing as a mode of life and the influence of Hadot's works, in which this tradition was reconstructed, revived, and integrated with a modernized conception of spiritual exercises. But these connections and influences are less significant; the point now is about the direct borrowing of the generative principle of the *whole* conception, together with a set of ideas accompanying it. The generative principle

---

114. Hadot, *Spiritual Exercises*, p. 301.

of the practice of the self is a specific principle of the constitution of the subject: this constitution is presented as a process of self-transformation in which the subject becomes himself (his "true self" and so on). In Kierkegaard's *Either/Or* the central theme is precisely a new principle of the constitution of the subject: the constitution of the self as a process of self-transformation in which the subject becomes himself (his "true self" and so on). Here this process is called the "choice of oneself" or, more precisely, the "ethical" or "absolute" choice of oneself. This is how Kierkegaard describes the principle of the constitution of the self in ethical choice: "He who chooses himself ethically has himself as his task.... By the individual's intercourse with himself he impregnates himself and brings himself to birth.... The great thing is not to be this or that but to be oneself, and this everyone can be if he wills it."[115] This is how Foucault describes the principle of the constitution of the self in the practice of the self: "The goal of the practice of the self is I myself.... An individual gradually transforms himself into himself.... That is the final goal for every individual" (146, 28, 147R; 122, 18, 123F). One can find great many similar formulas in both authors: since these formulas express the central principle of their theories, they are repeated many times and in many variations. There is no doubt that the two philosophers propound the same principle. Its development is a specific philosophical discourse: the discourse of the "self" constituted in the unfolding of its relation to the same "self." It is introduced into philosophy by Kierkegaard and is personally associated with him (in contrast to the general existential theme of philosophy as the art of living, which one finds in a great many thinkers). Thus I would consider it as proven that Foucault has borrowed this principle and discourse from Kierkegaard.

As I have pointed out, Foucault's conception also "silently borrows" from Kierkegaard the set of basic ideas accompanying the central principle. Let me indicate two such ideas, which are of prime importance in this conception. As Foucault himself acknowledged, his paradigm of conversion to the self bears the influence of Hadot, to whom also belongs the contemporary elaboration of the concept of *epistrophē* as well as the *epistrophē-metanoia* opposition. But this paradigm is present in Kierkegaard too. Though not concerned with its ancient history, he nonetheless insistently asserts in almost all his texts the principle of concentration on one's own inner reality, a principle that is virtually identical with conversion to the self. Inner reality, *Inderlighed*, one of the quasi-synonyms of the "self," is one

---

115. Kierkegaard, *Either-Or*, pp. 262, 263, 181.

of Kierkegaard's key concepts; and a summoning to, a pathos of, conversion to one's inner reality is a leitmotif of his philosophy. Taking this into account, one of Hadot's objections to Foucault acquires primary importance: his conception is "too insistently concentrated on the 'self' or at least on a certain conception of the 'self.'"[116] As is clear from this, Hadot does not share the position of one's concentration on the "self" forcefully advocated by Foucault. On the other hand, such concentration is the most characteristic feature of Kierkegaard's thought.

Now the next point: As we recall, the most important role in the conception of the practices of the self is played by the comparison and contrasting of the central principle of the self-transformation of the self with the principle of "know yourself," *gnōthi seauton*. Such comparison and contrasting is one more conceptual line initiated personally by Kierkegaard, who carefully collates the central principle of the (ethical) choice of oneself with the principle of "know yourself," and the relation he establishes between them is extremely close to Foucault's relationship between the principles of care of the self and knowledge of the self. "The expression *gnōthi seauton* has been repeated often enough and in it has been seen the goal of all human endeavor . . . but it cannot be the goal if it is not at the same time the beginning. The ethical individual knows himself, but this knowledge is not a mere contemplation . . . it is a reflection upon himself which itself is an action, and therefore I have deliberately preferred to use the expression 'to choose oneself' instead of know oneself. . . . This self which the individual knows is at once the actual self and the ideal self . . . to which he has to form himself."[117] Here the principle of knowledge of the self becomes just a part of man's tasks; it gets included in the paradigm of transformation of the self. We find exactly the same thing in Foucault.

The next instance of Foucault's close connection with Kierkegaard is of a different character. Foucault's characterization of the Christian model of the practices of the self is centered on the thesis of complete self-renunciation *(renonciation)* as the goal and the sense of Christian ascesis, the guiding principle of Christian consciousness and Christian faith. Self-renunciation, too, is one of Kierkegaard's concepts, one of the fundamental concepts in the analysis of the believing consciousness that he carries out in *Fear and Trembling*. But virtually the main idea of this famous book is the distinction the author makes between self-renunciation and faith!

116. Hadot, *Spiritual Exercises*, p. 300.
117. Kierkegaard, *Either-Or*, p. 263.

The Dane presents this distinction in many ways, including even an allegorical form: he depicts two figures, the "Knight of Self-renunciation"[118] and the "Knight of Faith," and describes their differences in detail. Self-renunciation, according to Kierkegaard, is not opposite to faith, but "the last stage before faith"; it is necessary for faith but does not have the latter's principal and decisive element, which is described as the unshakable conviction that the beloved thing surrendered in self-renunciation will be reacquired in some unfathomable and "absurd" manner — so that in the last analysis the way of faith leads not to (self-)devastation and destruction but to the new and only true fullness of personhood. As for Foucault, he never proves his thesis (I shall discuss it again in Section II). He avoids talking about the *telos* of the Christian practices of the self, and he does not conduct his own analysis of the higher stages of their way, an analysis which would demonstrate that at the end of this way one truly finds self-renunciation and self-destruction. Surely, Foucault knows perfectly well (as every student-philosopher knows), that Kierkegaard's most famous text makes a cardinal distinction between self-renunciation and faith. But he nonetheless keeps repeating — just as insistently but with just as little proof — that the end of the Christian practice of the self is self-renunciation. With this he defiantly contradicts Kierkegaard, without even mentioning his name.

We find a similar treatment of Kierkegaard also in another well-known theme of his philosophy, the widely popular schema of the three "stages on life's way," or the stages of the development of consciousness, which, according to Kierkegaard, progresses from an esthetic formation to an ethical and then religious one, to faith, in which alone can the fullness of the constitution of the subject be attained. Foucault even fleetingly refers to this schema in the 1982 Course, when he speaks of "the individual frozen at the esthetic stage" (this is the sole semi-explicit mention of Kierkegaard in Foucault's works). This schema is firmly present in Foucault's consciousness, especially in the final period, when his thought is dominated by the collision between the esthetic and ethical consciousness, by the problem of "morality" (let us recall the title of his last interview). In the philosophical aspect, the goal of the "esthetics of existence" is exactly the resolution of this collision. What is the nature of this resolution, if one compares it with Kierkegaard's schema? The religious formation of consciousness is wholly discarded, as destructive of the "self." The ethical formation is identified

---

118. In English translation this figure is usually called the "Knight of Infinite Resignation." — Trans.

with the esthetic one and is absorbed into the latter: there is no ethics except esthetics. Meanwhile, the esthetic formation, in the form of the esthetics of existence, turns out, in sum, to be not just supreme but even all-engulfing and unique. The dependence on Kierkegaard is concealed in this resolution too; it arises in the framework of his general picture of the three formations of consciousness. But it is clear that what we have is a direct and sharp negation of Kierkegaard's schema.

In the two last points that I have discussed, a certain pattern or stereotype emerges: the stereotype of taking some elements of Kierkegaard's philosophy and exploiting them after subjecting them to sharp distortion and/or diametrical reversal. This stereotype was apparently typical of the way Foucault treated his predecessors' ideas, and it was even noticed by one of them. Louis Althusser, one of his teachers of philosophy, writes the following: "In the stream of [Foucault's] ideas and under his pen the meaning of expressions borrowed from me was transformed into its diametrical opposite" (as quoted in Eribon, *Michel Foucault*, p. 83). In the light of this remark, one can suppose that it is this trait of his that Foucault has in mind when he talks about his attitude toward different philosophers: "There exist for me three categories of philosophers: Philosophers I do not know; philosophers I know and talk about; philosophers I know and do not talk about."[119]

If my supposition is correct, in Foucault's world both Kierkegaard and Althusser belong to the third category, although it is characterized with great moderation: more exactly, it includes authors to whose ideas Foucault imparts an opposite meaning when he borrows them.

In the light of this stereotype we must look more attentively at the first point, in which we described a seemingly simple borrowing of the principle of "the transformation of the self into the true self." Of course, here too one can find distortions and reversals. For Kierkegaard the principle "to become oneself" is a principle of the transcendent constitution of the subject, whereas for Foucault it is a principle of the entirely immanent constitution of the subject. Both philosophers compare Stoic and Christian conceptions. For Kierkegaard the Stoic conceptions give an example where the "ethical choice of the self" is not attained (but only an "abstract cultivation of virtues"), whereas true ethical choice is organically connected with the structures of the Christian consciousness and, first and foremost, with repentance, tied closely with the practices of confession *(l'aveu)*. In Foucault

---

119. Foucault, "Le retour de la morale," *DE II*, no. 354, p. 1522.

we find the opposite: in its mature and perfect form "the transformation of the self into the true self" is realized, first and foremost, in the Stoic practices. This is where we find the most radical divergence. The concepts and paradigms put forth by Kierkegaard achieve their perfect realization in the Christian consciousness, whereas they achieve imperfect, defective, or distorted realizations in the pagan consciousness. Foucault takes over these concepts and paradigms, but he omits any reference to Kierkegaard. Furthermore, he gives a completely opposite twist to the Kierkegaardian interpretation and presents the practices of the self as having found perfect realization in the pagan consciousness and irreparably defective realization in the Christian consciousness.

I don't know why Foucault acted in this way, and I prefer to avoid any guesswork. But I spent much time in Kierkegaard's world, and I know that it is not the world of abstract, academic thought. It is a living organism, living an intense life, religious and existential. (See my detailed description of this life in chapter 7 of the Russian *Diogenes' Lantern*.) In Foucault's treatment, some basic organs of this organism, alienated, are forced to serve ends diametrically opposite to those which inspired all its life, and this treatment hurts it.

11

Spiritual Practice, Synergic Anthropology, and Foucault's Project

"Foucault was, among other things, the leading exponent of ascetic values and concepts among major intellectuals of our time."[1]

In this section I intend to compare Foucault's conception of the practices of the self with the project of synergic anthropology. In the large, the two projects have many similar traits: they develop a nonclassical anthropology, abandoning the essentialist anthropological model of Aristotle-Descartes-Kant and putting forth the alternative idea of the constitution of man in certain practices of self-transformation. In addition, both projects include the study of practices of Christian ascesis, thereby proposing two different conceptualizations and hermeneutics of this sphere of anthropological experience. It is clear from this that the task in this chapter is a twofold one: to survey the general foundations, principles, and structures of the two projects in their differences and similarities; and then to compare the two relevant treatments of ascetic experience in order to establish which of them is more adequate.

I will begin this comparative survey with the general context and genesis of the two projects. First on Foucault's project: In what context, in what logic of ideas does the conception of the practices of the self arise in Foucault's work? He reveals this in the Introduction to the second volume of *The History of Sexuality*, which serves as a kind of explanatory note to

---

1. Geoffrey Galt Harpham, "Old Water in New Bottles: The Contemporary Prospects for the Study of Asceticism," *Semeia* 58 (1992): 140.

his entire later period. The initial goal of *The History of Sexuality* was to study sexuality in the framework of Modern Times and of previous orientations centered entirely on discourses of power and knowledge, but with further reflection Foucault went far beyond this framework. What became most important for him was the question of the coupling of sexuality and morality, the question why a rich and complex "moral problematization" forms around the phenomenon of sexuality. It is this reflection on moral problematization that causes a shift in the frame of the project. The problem field of his studies changed and became oriented to the subject, to problems of "the relationship of self with self and the forming of oneself as a subject," while the historical base of his studies shifted "from the modern era . . . through Christianity to antiquity," inasmuch as "in order to get a better idea of the forms of the relation to the self, I was compelled to go farther and farther back in time." Here the subject emerged as the problematizing subject, as seeing himself, his actions, and the world of his existence as a "problem," as an element of questionability. And this view of the subject led to yet another object of study, the last and most important one. Moral problematization, taking place in the subject's consciousness, exerts a formative influence on a certain genre of practices of the subject: on those practices the subject directs at itself and in which it purposively changes itself. These are precisely the practices of the self. It is exactly by way of establishing the logical connection of these practices with moral problematization that Foucault comes to their first detailed definition: "This problematization was linked to a group of practices . . . by which men not only set themselves rules of conduct but also seek to transform themselves, to change themselves in their singular being."[2] That is the logic of the genesis of the conception: this logic leads from the moral problematization of sexuality to the practices of the self, the study of which is immediately understood as a study of the constitution of the subject and the hermeneutics of the subject.

The roots of the second project under examination here, synergic anthropology, lie in a wholly different soil and context. The principal concept describing this context is the *Eastern Christian discourse,* the specific discourse created by Eastern Christianity (Orthodoxy) and constituting the spiritual, conceptual, and epistemological fund for forming the Eastern Christian (first Byzantine and then Russian) mentality and cultural-civilizational organism. The structure of this discourse is characterized by

---

2. Michel Foucault, *The Use of Pleasure: The History of Sexuality,* vol. 2 (Harmondsworth and New York: Viking, 1985), pp. 6, 9-10.

the presence of a generative core consisting in a specific type of experience: the authentically Christian experience of striving toward Christ and union with him, the experience of Christocentric communion with God, affirmed as constitutive for man and forming his personhood and identity. The cultural-civilizational organism contains a special sphere where this kind of experience is cultivated: this is the hesychast mystical-ascetic tradition, which creates and sustains the ensemble of practices producing the experience sought for.

With respect to philosophical thought the Eastern Christian discourse is its initial all-inclusive context defining the self-identity of this thought as well as its deep-seated tasks and strivings, its guiding intuitions, its customary practices and stable patterns, its typology. However, in the Eastern Christian tradition, philosophizing always occurs not in a single context but in an intersection of contexts. For the philosophical consciousness, philosophy as such is represented by the European philosophical tradition: the latter is viewed as the "home of philosophy" and it provides the language for philosophy. Therefore the European philosophical process is the other inevitable and necessary context for philosophical thought, and we find its most immediate and direct context in the results of this process at the present moment: in the current philosophical situation, in its existing concepts and problem field.

After the complex peripeteias of the development of Russian thought in the twentieth century, characterized by a tangled relationship between its European and Eastern Christian contexts as well as between its philosophical and theological discourses, and given the restoration, "after the interruption,"[3] of the possibility of free philosophizing in Russia, there arose the undoubted necessity of *"another beginning,"*[4] of a new reflection of philosophical thought on its own sources and foundations, on its own "two-context" (European and Eastern Christian) nature. First of all, this presupposes reflection on the Eastern Christian discourse itself, on its structure and the structure of its core; and it is clear that the renewal of the understanding of this core is the key to the whole problem of *another beginning*. Thus the most essential and cardinal task consisted in *the mod-*

---

3. The state and prospects of Russian philosophy at the moment of the collapse of the Soviet regime are discussed in my book *Posle pereryva* (After the Interruption) (St. Petersburg: Aleteja, 1994).

4. "Another beginning" is the Heideggerian concept expressing the "archaeological nature" of philosophical thought: the permanent need for the latter to renew its connection with its beginnings or sources, *archai*.

ern reconstruction of the originating experience of the Eastern Christian discourse — of the anthropological and meta-anthropological experience of the practices of the Christocentric communion with God. Insofar as this statement of the task entailed the affirmation of a certain kind of extreme (or more precisely ontologically extreme) experience as man's constitutive experience and presupposed a systematic reconstruction of the practices cultivating it, then in the European perspective, in the prism of the current philosophical situation, it meant an approach that discarded the essentialist Aristotelian foundations of the anthropological discourse and developed a representation of man in an active and energic discourse, in the being-action dimension. It was thus oriented toward the construction of a certain "anthropology of practices," or "energic anthropology."

II.1. Reconstruction of the Hesychast Practice

The reconstruction of hesychast experience and practice, a reconstruction carried out in the form of their phenomenological description, is part of synergic anthropology, forming its foundation. Already at this stage synergic anthropology starts to formulate a certain paradigm of the constitution of the subject and develops a particular hermeneutics of the subject. Their character is determined by the specific nature of the practice studied. In analyzing the hesychast practice we find in it a very special kind of anthropological phenomenon: an example of a practice which, in a certain sense, is alternative with respect to all practices and strategies developed by man in his usual, ordinary existence. This quality of alternativeness is already implanted in the property noted above: in contrast to all "practices of ordinary existence," the hesychast practice is not only an anthropological practice but also has a meta-anthropological dimension. This is expressed in the fact that the hesychast practice is strictly goal-oriented, although its goal is not present anywhere in the horizon of empirical being: the goal is ontologically beyond this horizon since it is Divine being — the different ontological horizon, or the "Other-being," as I will call it. For such a nonpresent goal, or "trans-goal," the term *telos* is appropriate. Orientation toward the *telos* of Other-being is the defining property that makes hesychast practice an alternative practice and imparts to it an extremely specific nature: one that requires special conditions for its realization, makes the practice a process with a type of dynamics unique in anthropology, akin to processes of self-organization, and also imparts to it the character of the total self-transformation of the human being. The latter characteristic means that hesychast practice is a practice of the self, but a very special, *ontological* practice, possessing many essential features not described in Foucault's theory.

The reconstruction of such a specific anthropological phenomenon requires methodological reflection. It is clear, first of all, that the completeness and reliability of the phenomenal base that is used are particularly essential here. Therefore, synergic anthropology defines and describes the complete field of hesychast experience; that is, it reconstructs the complete chronotope of the ascetic tradition — all the stages of its development and all the regions of its dissemination. Analogously, it describes the whole body of sources and analyzes them, revealing the specific features of the ascetic discourse and establishing the correct rules for reading it. This entire preparatory base is presented in my book *Toward a Phenomenology of Ascesis*, where the reconstruction of hesychast practice is described in all its aspects. In the present work I examine only those points of this reconstruction that are necessary for comparing my concept of synergic anthropology with Foucault's practices of the self: in other words, in what follows I analyze only those points of the ascetic tradition that reveal the key features of the structure and nature of the hesychast phenomenon.

First and foremost is the feature that makes us assign hesychast practice and similar practices to a special class of anthropological phenomena. The ontological practice of the self, oriented toward a meta-anthropological *telos* (I will call this type of practices *spiritual practices*), is characterized by the necessary presence of an *organon* of its experience. The term "organon" has here its precise Aristotelian meaning: for a specific type of experience, the organon is its complete practical-theoretical canon — a code of rules for the organization, testing, and interpretation of experience. Experience that possesses the organon is a "complete experience," i.e., one that completely explicates a certain nature, *physis*. Hesychast experience refers to man's nature, and hence the fact that it possesses the organon signifies that hesychast practice is constitutive for man. It also signifies that hesychast experience does not simply follow a specific method but represents an experience that is fully tested and arranged, that is processed and reflected on (albeit not in a philosophical discourse). And so, hesychast practice is a spiritual-anthropological process that possesses the organon of its experience and forms up man's constitution, the actualization and fulfillment of his nature. The supremely nontrivial fact that hesychast practice works out the genuine and full-fledged organon of its experience, is explained by nothing other than the ontological status of its *telos*. It is possible to advance toward such a *telos*, which is surely absent in the horizon of empirical experience, only if one has a precise and complete "travel instruction," which is exactly what the organon of the practice's experience is. The creation of the organon as well as its application and preservation represent a special and complex activity, which cannot be accomplished by any individual adept of the practice, but requires a certain community. And so, it turns out immediately that spiritual practice is an anthropological, an individual practice that necessarily includes transindividual (intersubjective, collective, or "communal") dimensions. I will come to the communal dimension of spiritual practices later.

The next key property defines the general structure of the spiritual-anthropological process of the practice: this structure corresponds to a stepwise paradigm or, to put it more simply, it has the form of a "ladder." This is a purely experiential fact: quite early in its development the hesychast practice, as well as every other tradition of spiritual practices, discovers that progress toward the meta-anthropological *telos* has the character of an ascending process, which is divided into clearly delineated steps. What is the character of these steps? The process of the hesychast practice is an ascending progress toward the Other-being, toward communion and union with Christ, by means of "holistic self-transformation," a human being's total transformation of himself. An important complement must be made, however: an individual transforms himself here not in his material composition but *in his energies;* and the farther he progresses, the more this transformation is realized not by his own energies but by the energies of the Other-being. In the process of the practice the Other-being manifests itself as a "Source-beyond-there," as the source of certain energies that take part in the transformation, but are perceived by the individual as not belonging to him. His own energies reach a contact with these "other energies" at a certain stage of the practice, and the perfect union of these two ontologically different energies is the finale of the practices, their *telos*, a union conveyed by the theological concept of *deification, theosis*. This complement consists of two parts; the second part concerning the Source-beyond-there and its energies will be discussed a bit later. As for the first part, it tells us that in spiritual practice an individual emerges as an "energic formation": he treats himself as a combination of various energies, spiritual, psychical, and physical, and he successively transforms this formation. Thus, on the "ladder" of the practice each of the steps corresponds to a certain specific configuration (or mode of organization, or regime of activity) of the set of all the human being's energies.

Spiritual Practice, Synergic Anthropology, and Foucault's Project

Apart from the individual steps, one can discern in the "ladder" of the practice some large structure — groups of steps having not very different energic configurations and playing similar roles in the process. This structure makes it possible to clearly see the general course of the process and the logical order of anthropological tasks that it states and fulfills. My reconstruction identifies three such groups, thus dividing the whole process of the practice into three big parts, or "blocks." In the first of them, the practice establishes itself as an *alternative* anthropological strategy; in other words, at its steps, an individual performs the going-out of his ordinary existence, of the "world." (This important ascetic concept has not a substantial but an energic meaning, signifying a specific mode of man's existence and man's organization; it can also be conceived as the ensemble of all anthropological strategies corresponding to ordinary existence which is subordinate to the social empirical being.) This "going-out" is a dramatic act realized by means of extraordinary and radical practices that shake, revoke, or destroy all the previous stereotypes of an individual's consciousness and existence. It is struggle and combat, and therefore after the initial block, one must "come to one's senses"; it is followed by the step of inner quietude and solitary concentration: this is *hesychia*, which has given its name to the entirety of the Eastern Christian spiritual practice. The fruit of this block is a reorientation of the vector of an individual's principal efforts; thanks to this reorientation, there can begin a new construction of oneself, in which an individual's energies are organized into such configurations that are in no way possible in secular strategies.

In the next block, the central one, a key and cardinal problem is solved: the problem of creating an "ontological mover," an anthropological dynamics that would make actual progress to the Other-being possible, imparting to the process the character of an actual ontological transformation of anthropological reality. But the ontological transformation can take place only through energies "from beyond," not through an individual's own energies. Thus a necessary precondition for such dynamics is the openness of the "energic human being" to the energies of the Other-being, the possibility that his energies will encounter them and that the two ontologically different energies will reach mutual coordination and harmony. The attainment of this openness or (which is the same thing) *unlocking of the human being* is the key and decisive moment of the spiritual-anthropological process. The encounter and harmonious coordination of the energies of the individual and the energies of the Other-being (the Divine energies), of these energies that differ ontologically, is called *synergy* in Byzantine theology.

The following stages, at which the dynamics of the ontological progress already exists and is manifested in action, form the concluding block of the process. Generated by the Source-beyond-there, this dynamics is a specific product and property of the spiritual practice, and owing to this dynamics the practice at its highest stages clearly displays alternativeness to all "ordinary," "horizontal" anthropological strategies. The closer it approaches the *telos*, the process of ascent on the steps more and more acquires the character of spontaneous form-building, reminding one of self-organization processes and synergetic processes in which spontaneous generation of a hierarchy of dynamic structures occurs. An individual's energies are spontaneously or "automatically" organized into new forms that are structurally higher than and represent a development and surpassing of the preceding forms. The human being acquires new capabilities in these new forms, which can be called the *self-surpassing of the individual* (the formula used by hesychasts to express the essence of their practice). The steps themselves, or the anthropological energic forms, represent configurations of human energy that cannot be realized separately from the process of the practice. All the fullness of the ontological transformation is not achieved within the limits of this process; however, the abundant experience of all spiritual practices attests to the existence of a particular zone of actual approach to the *telos*. What distinguishes this zone are explicit signs of an incipient transformation of an individual in his fundamental structure, signs of a change of the fundamental predicates of the mode of human existence. These signs appear, in the first place, in the sphere of man's perceptual modalities and represent manifestations of a certain new perceptual modality of a radically different nature. This new modality, which in hesychasm is called "noetic senses" *(noēta aisthēsis)* is characterized by such qualities as "synesthesia" (the new modality is a unified synthetic capability, not divided into particular senses) and "panesthesia" (the new modality is a capability belonging to the entire human being as a whole, without any connection to particular organs). In the experience of these radical perceptual transformations, in which the fundamental predicates of the mode of man's being begin to change, the meta-anthropological dimension of spiritual practice is actualized. It is necessary, however, to underscore that the phenomena occurring in the zone of approach to the *telos* do not find a completely adequate description in the objectified language of systems and processes; here this language must be combined — according to the complementarity principle in the sense of Bohr — with the discourse of personal communion (communion with God).

In each of the blocks described there occurs a nontrivial anthropological construction: for virtually each of the steps of the ladder, special technologies of the self develop. In this respect the initial block is the richest one thanks to the diversity of problems it solves. "The going out of the world" includes two major tasks: first, the task of performing the act of choice, of making and actualizing the fundamental decision as to the selection of an alternative, "vertical" strategy, or (which is the same thing) of passing through the "Spiritual Gates," which the initial step of the spiritual practice is often called; secondly, the task of overcoming the obstacles — the "passions" — that appear immediately on the other side of the Gates and block the path of ascent. Entering onto the path of the practice is a complex anthropological and ontological event whose structure is well conveyed by the metaphor of the Gates: the Gates divide two regions different in some respects, and the individual entering strives, first, to leave one of them and, secondly, to integrate himself in the other. In this dual structure the first element is the event of departure, of a rupture with the former world, the event of a-version from the latter and con-version towards a new world — to another strategy, to another mode of existence (we remind the reader that the "world" is understood here in an active-energic voice as a world of strategies, practices, energic forms). The second element is the event of going into and integrating oneself in this other mode, the appropriation of its principles and values. The first element is called *Conversion*; the second element is called *Repentance*. Together, they form a fundamental event of change of consciousness, a "transformation of the mind," which corresponds to the Greek term *metanoia* and constitutes the content of the initial step of the spiritual practice. Thus, owing to the ontological and alternative nature of the spiritual practice, its initial step acquires the following structure:

Metanoia = Conversion + Repentance.

The phenomenon of spiritual practice and the anthropological paradigm connected with it have to this point escaped the attention of scholars, and therefore this structure has not been noticed and described, although its two elements have often been studied separately. Also, the specific form in which Conversion appears in the "transformation of the mind" has gone virtually unnoticed. As the initiatory event of the spiritual practice, Conversion does not in any way correspond to the ancient model of conversion-return, *epistrophē* (in accordance with which, as we have seen, Conversion is treated by both Foucault and Hadot). Here it is placed in a totally different conceptual context, implied by the ontological structure of the practice, which, in turn, is determined by the presence of the Source-beyond-there. As sufficiently profound discussions of the spiritual practice always underscore, by its nature its beginning has likeness to its end, its *telos*: its conceptual structure, too, is necessarily connected with the Source-beyond-there. In man's ontological situation the presence of this Source is expressed as an ever-present presupposition, as an open possibility of union with it — and in this sense as its *call* which is present in reality and may be heard or not heard by an individual. And Conversion is when an individual "turns around to hear this call," when he catches it and recognizes it as a call from the Source-beyond-there, and responds to it wholly, with his whole being.[5] By virtue of the fact that the Caller is located in the ontological beyond, this Conversion, or Turning, to the Call brings with it what is expected of the Christian conversion: a choice of strategy alternative to the whole ordinary mode of existence. It also signifies, like Conversion in Foucault's treatment, conversion to the self, inasmuch as the alternative strategy entails a total self-transformation (and one that is incomparably more radical and global than in the Hellenistic practices).

Further, Repentance is necessary if one is to establish oneself firmly in the chosen alternative mode of existence. Its character is defined by the fact that in this case the "alternative mode" is *ontologically* alternative, that is, alternative to the former mode in the strongest possible sense. Therefore all of its positions, principles, and assessments of reality are different from those in the former mode and reject them radically; in order to enter into the alternative mode of existence it is necessary to decisively renounce the old mode with all its contents. The hesychast practice realizes such renunciation in a rich ensemble of very diverse technologies of the self, all of which, by their character and purpose, contradict the unconverted reason and have always, in all epochs, evoked misunderstanding and negative reactions. We have observed such reactions in Foucault too (by the way, the object of his most acute aversion, the "practices of confession," are by no means the most shocking and extreme ones in the arsenal of the technologies of repentance, which includes stylitism, severe fasting, the vow of silence, tears of many kinds, and so on).

5. A treatment of Conversion in spiritual practice on the basis of an ontological analytic of Call and Response is given in my works; see "Pravoslavnoe pokayanie kak antropologicheskij fenomen" (Orthodox Repentance as an Anthropological Phenomenon), in *Issledovaniya po isikhastskoj traditsii* (Studies in the Hesychast Tradition), vol. 2 (St. Petersburg: Russkaya Khristianskaya Gumanitarnaya Akademiya, 2012) [in Russian].

Spiritual Practice, Synergic Anthropology, and Foucault's Project

It would seem that this initial block of the spiritual practice is precisely the part of it that confirms Foucault's view, in that it effects "destruction of the self," "renunciation of the self," and so on. In ascesis, repentance truly is a "renunciation of the self," a pitiless examination of all the foundations and contents of the self, a thorough discarding of the entire former order of one's consciousness and existence. Nevertheless, Foucault's view is categorically incorrect, for the meaning and the content of the integral phenomenon of the spiritual practice are directly opposed. To consider the technologies of repentance as something separate and autonomous is an absurd position that robs them of their meaning: repentance is exclusively *the beginning of a spiritual process*, the process whose goal and meaning lie not in destruction but in construction. When describing the higher steps of the practice, I already observed that on these steps the practice acquires the character of the spontaneous generation and organization of new, more perfect anthropological energic forms; and soon I shall show that this expresses the creative nature of the spiritual practice — as a practice in which a new type of subjectness, a new paradigm of human constitution, is discovered and actualized. The source of this creativity of the spiritual practice is the Source-beyond-there, the union with the energies of which represents the ontological dimension of the practice. This dimension is manifested most clearly at the higher steps, but its role is significant at any of them. A strong emphasis on the ontological nature and function of repentance, on strict necessity of the latter both at the beginning and in the whole path of ascent to deification, is the most distinctive characteristic of the hesychast practice and of the Eastern Christian discourse.[6] It is exclusively this ontological nature of the phenomenon that grounds and explains all the extreme aspects of penitent self-renunciation and self-destruction.

There is a necessary relation, a universal law, if you will, of the practices of the self: to the extent the *telos* of the practice is not purely immanent, to that extent the initiation of the practice requires "self-renunciation" and a break with the ordinary mode of existence. *The transcendence of the telos of the practice implies the "alternativeness" of the path of the practice* and is manifested in the initiation of the practice by the presence of an entry barrier that is overcome by special "preparatory" technologies of the self which

---

6. Western Christianity has long had a different, somewhat more superficial or "take-it-easy" attitude toward repentance: it did not connect any ontological aspect with it, but has instead regarded it chiefly as a juridical function implementing the "restoration of relations" between the penitent sinner and God and as an "economic" function involving the maintenance of a certain stable order of religious life.

can be called "technologies of purification" (with this term understood in a broad and generalized sense). In the Hellenistic practices examined by Foucault he asserts (though Hadot denies) the total immanence of the *telos*; and accordingly he does not find in them any technologies of purification: the structure of the Gates of these practices is only a conversion without any purification, repentance, etc. In the broad spectrum of the other practices of the Ancient World — the mysteries, the Gnostic and Neoplatonic practices, etc. — some degree of transcendence was usually presupposed, even if often it was just vaguely articulated; and accordingly the Gates of these practices included practices of purification. Finally, in Christianity the transcendence of the *telos* of the practice reaches the absolute extreme, an ontological rupture; and accordingly, at the beginning of its path, one finds unprecedented practices of repentance — a kind of "extreme purification," requiring a complete departure from the world.

In sum, we arrive at something extremely simple and generally known: renunciation of one's "old self" in repentance, at the Gates of the practice, is a necessary premise for acquiring one's "true self" at the end of the practice. The "true self," built-up in the practice and in no way reminiscent of the total "destruction of the self," is not a declaration or fiction but a content of experience that has been verified, described, and profoundly explained many times. But all of his talent as a philosopher could not overcome Foucault's furious nature, for which it was impossible to recognize this experience. Two conceptions of self-creation collided here: for Foucault such creation consists in the esthetic cultivation of the untouchable initial core of the "self," whereas for Christianity it consists in self-unlocking and person-creation (based on an unprecedented fullness of self-disclosure and self-revision). These two conceptions are mutually exclusive. For Foucault, as we see, the path of Christianity is one of self-destruction, not one of self-creation, but it must be said that, for Christianity, Foucault's path is also not self-creation at all. According to Christianity, what lies in the "untouchable initial core" of the "self" includes necessarily man's passions,[7] and the cultivation of them is not creation but the obedient service of them, enslavement by them.

The concluding task of the initial block of the hesychast practice consists in the struggle with passions, or the "Invisible Warfare." Even though there are only eight principal "spirits of evil," man has a multitude of pas-

---

7. Psychoanalysis views the core of the "self" in a rather similar way, which is why there was always an element of restrained unacceptance in Foucault's attitude toward it.

sions. The tasks of struggle with them have been examined separately for each passion and collectively for the whole regiment of them; and as a result the elaborated repertoire of anthropological, psychological, and intellectual methods, know-how, and technologies of the self far surpasses, at least numerically, all the systems of spiritual exercises in the ancient practices. There is no need for us to examine these technologies, but it is worth indicating the general motivation for the Invisible Warfare. The phenomenon of passion is interpreted in terms of process and dynamics: passion is an "unnatural state" (as Isaac the Syrian puts it) in which an individual has become enslaved by some impulse of his, has fallen into total subservience to it, and, endlessly reproducing his fixation on it, is deprived of freedom of action and development. In particular, he is incapable of spiritual ascent. Therefore the struggle with passions and the attainment of their eradication, dispassion *(apatheia)*, is release from captivity to them, liberation. Here, in the principle of the expulsion of passions as well as in the extolling of *apatheia*, there is no need to deny closeness to and some direct borrowings from the Stoics. But the context and the role of all this work in the process is important: if, in Stoicism, dispassion, together with *ataraxy*, constitutes the end of the practice of the self, then, in hesychasm, dispassion is only the end of the "Praxis," of the "active" part of the process, after which follow the higher steps, the "Theoria." In the two traditions, Christian Hesychasm and Stoicism, very different configurations of energies are called *apatheia*. Sending the reader to my book *Toward a Phenomenology of Ascesis* for a detailed analysis of hesychast dispassion, I will only note briefly that this dispassion, in contrast to the Stoic type, is "dithyrambic," two-directional: in relation to the things of the world, to the passions of the soul, it is an immobility, whereas in relation to the *telos* of the practice, to the highest steps of the ladder, it is an intense vital striving.

The specific technologies of the central block of the practice begin with the formation of a special modus of consciousness: the *modus of watchfulness*, which seems to me one of the most important anthropological discoveries of the spiritual practice (it is particularly well developed in hesychasm, but it is necessarily present in other traditions as well). In the early period, the principal activity that the ascetic consciousness has to sustain and to which its other activities have to be subordinated was *diakrisis* — discernment, discrimination. John Cassian puts it at the center, devoting the second of his *Conferences* to it and tracing its primacy back to Anthony the Great himself, the founder of ascesis. It includes, first of all, the principle of measure and moderation and is clearly consonant with Stoic thought;

Foucault justly remarks that this concept was already actively introduced into the practice of the self by Epictetus. Experience showed, however, that the principal activity or modality of the ascetic consciousness had to be something else. The ontological nature of the Christian practice is manifested in a change in the character of the activity and modes of operation of the consciousness compared with pagan practices. Thus at the next stage, in the fifth to seventh centuries, a painstaking search leads to the gradual fixation of a new central concept — *nepsis* — in the ascetic discourse. It is sobriety, but in the ascetic context, the better translation is watchfulness. Russian hesychasm even created a neologism, *trezvenie*, for this concept, which was very different in meaning from the ordinary word for sobriety, *trezvost'*, the antonym of drunkenness. The change is not limited to this: together with watchfulness, a number of related concepts become very significant, and collectively they form a unified conceptual structure that characterizes a particular mode of organization and operation of the consciousness; and it is to this mode that we give the name "modus of watchfulness."

Watchfulness can be defined as the principle of the concentrated and sharply directed vigilance of the consciousness; it becomes the *topos* of the consciousness, the generative principle of its particular modus, by virtue of the fact that it brings the consciousness into a specific order in which the opposition between passivity and activity is overcome. The rich and subtle discourse describing the work of consciousness in the modus of watchfulness was created in the period of so-called "Sinai hesychasm" (seventh to tenth centuries). The basic concepts of the modus include attention (subdivided into different types: attention of the mind, of the heart, etc.), memory (also divided into several types), guarding of the mind/heart, self-observation, vigilance, discernment (no longer in the dominant position but by no means discarded), and so on. Ascetic discourse establishes connections between these concepts, and in particular it singles out in this modus its basic structure consisting of three elements: *watchfulness-attention-guarding*. Analysis of this modus in its operation leads to an important conclusion: consciousness in the modus of watchfulness is precisely what contemporary phenomenology calls intentional consciousness, and its activity represents an intentional act — an act of intellectual scrutiny in which the consciousness concentrates itself on a particular experiential content and focuses on it, purposively progressing toward a clear and distinct envisioning of it in all the dimensions and structures of its meaning. In the ascetic discourse watchfulness represents the direct analogue of intentionality.

Spiritual Practice, Synergic Anthropology, and Foucault's Project

The watchful, or intentional, consciousness is an essential element of the paradigm of the spiritual practice, an element that is not possessed by all the practices of the self examined by Foucault. But the essence and goal of the spiritual practice is ontological ascent, and in relation to it the modus of watchfulness plays only an instrumental and secondary, albeit important, role. The nucleus or core of the paradigm is a specific technology of the self (completely absent in Foucault's analysis): the coupling of the modus of watchfulness with a very different activity, which is precisely an activity of ascent as such. This is the famous *Attention-Prayer* coupling and it is considered to be the key to all the practice, so that its Greek form, *prosochē-proseuchē*, served as a kind of "hesychast mantra." In the central block of the practice the communion with God in prayer attains the form of unceasing prayer, which possesses the power to build up the higher steps of the practice, and this form is maintained by "attention," or more precisely by the modus of watchfulness, which through intentional focusing on it, vigilantly guards the space of the prayer against the intrusion of "thoughts" (in the sense of evil thoughts, *logismoi*). It is clear that this key technology of the self, demanding the simultaneous presence of two activities of consciousness that are radically different but closely harmonized, is unprecedented in its subtlety and difficulty. In this respect it is similar to another one of the central steps, which is its direct precondition: *the bringing of the mind into the heart*. This represents a unification and mutual coordination of man's intellectual and emotional energies as a result of which a cardinally new structure, the "mind-heart," is formed. By virtue of the joining, harmonization, and union of the two principal forms of human energies, this structure possesses a special strength and stability and is therefore used as the foundation for constructing further energic forms. According to Theophan the Recluse, it serves as "the lever by which one can set in motion one's entire inner world."

It is impossible to describe in a simple manner the subtle operations by which the ascetic's "mind" is brought into his "heart."[8] But more important for us than a detailed description is to observe that, in these technologies of the self developed in the central block of the spiritual practice, a certain new level of the organization of human consciousness and all human being emerges. When the consciousness organizes and maintains the operation of the *Attention-Prayer* coupling, when in a strictly defined manner it

---

8. See Sergey S. Horujy, *K fenomenologii askezy* (Toward a Phenomenology of Ascesis) (Moscow: Izd. Gumanitarnoj Literatury, 1998), pp. 105-9.

brings together and then unifies the intellectual and emotional energies, it actualizes simultaneously two of its different modi or aspects — say, in the first case, "intentional consciousness" and "mystical consciousness" — and purposively operates with them. Thus it discloses in itself a certain meta-level or meta-center, a kind of "higher authority," which has the capability and the power to govern nothing other than *the structure of the consciousness and the modes of its activity*! As Foucault says, the goal of the ancient practices was self-possession and self-control. Spiritual practice sets other goals, but here we see that in the course of this practice there is collaterally achieved such a degree of self-control compared to which the spiritual exercises of the Stoics look like the "self-control" of a toddler who knows already how to avoid tripping when he walks. The formation of the meta-level of consciousness was observed and reflected in hesychasm: Gregory Palamas called it the appearance in an individual of a "mind-bishop" (in Greek: *episkopos*, one who supervises or governs) which "prescribes laws for each force of the soul and each member of the body" — "laws" that arrange ascent to deification.[9] Thus this new (meta-)capability of human beings is by no means a goal in itself, but serves as a way to progress toward the *telos*; and the most essential thing is that its formation is not accomplished by human energies alone but only with the participation of energies from the Source-beyond-there.

Here we see that spiritual practice enters into a domain where it approaches synergy and directly becomes the practice of an individual's ontological unlocking. In the manifestations corresponding to these steps of the practice, an individual actualizes his relation to his ontological Other, and this relation is, by its very definition, constitutive for him. Accordingly, in this actualization he is endowed with a certain constitution whose character is determined by the *telos* of the practice. The *telos* of the hesychast practice is deification conceived as the perfect union of human energies with energies of the Other-being, Divine energies. Such union is actualized not analogously to the physical interaction of energies from two different sources but in the *paradigm of personal communion*. This paradigm has reference

---

9. It should be recalled here that the Stoics always had the idea of a unified governing center of the human being, a "guiding principle" *(to hēgemonikon)* which was usually identified with reason. In the light of this, the distance between the two traditions becomes completely clear: if you will, the "mind-bishop" is also a kind of "guiding principle," although not an immanent one, not an originally present one, but one that is acquired or formed. And it does not coincide with the mind but "bishops" over it — or more precisely it can be regarded as a mind elevated by God, by the Source-beyond-there, to the bishop's function.

to the concept of personality, and this is a fundamental distinction of the process. The *telos* of the hesychast practice and the human constitution that it determines are conceptualized in the discourse of personality; this is their principal and uneliminable feature.

"The personality is a discovery of Christianity," declares an old maxim whose modern interpretation and justification are presented, for example, in the works of Georges Florovsky. In this modern interpretation, that which is ontologically Other to the mode of man's being, i.e., Divine being, is treated as the ontological horizon of the being of the Person, or as "personal being-communion": the mode of being defined by the ontological (extratemporal) dynamics of unceasing and perfect interchange of being between the three Persons, or Hypostases. (The concept of Hypostasis, *hupostasis*, was formed by Cappodocian patristics on the basis of the concept of Aristotle's "first essence" through the radical "remelting" of the latter, as Florovsky calls it, and its identification with the hitherto completely nonphilosophical concept of "person-mask" or "person-role," *prosōpon*.) In accordance with the common dynamics that unites them, all three Hypostases have one common Essence, or *Ousia*. The extratemporal inter-Hypostatic interchange of being is called *perichoresis* (which signifies "going around a circle") and is regarded as the love that unites the Hypostases and as their communion. There is thus established the identity of three fundamental concepts: *perichoresis*-love-communion, each of which is understood as an ontological principle. Being mutually identical, these three ontological principles collectively characterize a certain ontological horizon, a mode of being, which by definition is precisely the being of the personality, or "personal being-communion" (in other words, this triple identity is manifested as the inner definition of the personality). Thus, the perfect personality (or Hypostasis) is formed in the perfect taking and perfect giving of being; and, evidently, in this mode of formation it also realizes perfect confirmation of itself to itself, perfect self-verification. By definition, this signifies that it is endowed with perfect *self-identity*. This conception of personality, developed in the Eastern Christian discourse, is known as the *theological or trinitarian personological paradigm*, in contrast to the *anthropological personological paradigm* developed in the West and expressing the principle of personality by means of the basic concepts "subject" and "individual." This paradigm implies that, in the horizon of empirical being, the fundamental triple identity is destroyed, *perichoresis* does not take place, and empirical man and his communion are a "personality" and "personal communion" only to a certain imperfect degree. As for spiritual practice, it is precisely that strategy of human beings through which they acquire genuine personality, are "created into persons" (to use Lev Karsavin's expression).

Indeed, the *telos* of the hesychast practice, Other-being, is the person conceived as personal being-communion and actualizing certain accomplished paradigms of "subjectness" (in the sense of a mode of human constitution) and self-identity. In synergy, or ontological unlocking, man's energies begin to conform with the Other-being's energies; in other words, they begin to enter and to be integrated into the order inherent to them. This means that, in the ontological unlocking of himself towards the Other-being, a human being partakes in the order of personal being-communion, in personhood. One can say that such partaking or participation occurs in the practice as a whole: the pre-synergic steps of the ladder lead to the unlocking, while its higher steps bring the unlocking close to its fullness, to deification. In the light of this, the concrete content of the paradigm of human constitution contained in the hesychast practice becomes clear: this is the paradigm of stepwise-increasing *energic participation in (trinitarian) personality and identity*, which, in turn, are defined by the ontological paradigm of *perichoresis*, the perfect interchange of being. The description of the spiritual practice as a process of the constitution of the human being, or "subject," in Foucault's conditional sense, i.e., of a subject formation, represents a certain "hermeneutics of the subject." In what follows, it will be unfolded and compared to Foucault's hermeneutics of the subject.

Being determined by the order of personal being-communion, the constitution of the human being and the hermeneutics of the subject in hesychasm and the Eastern Christian discourse in most diverse aspects exhibit their close connection with the phenomenon of personal communion and their dialogic nature. In this connection it is appropriate to make our terminology more precise: in the framework of the Eastern Christian discourse it is natural to give the name "personal communion" to the sort of dialogic communion that is oriented toward the order of "personal being-communion," i.e., of the perfect interchange of being, not of "information." Such communion implies a spontaneous expansion of the dialogic space, and has a tendency to go down to the ultimate depths, to the point of drawing into the space of communion all the fullness of the communing personal worlds. In spiritual practice an individual participates in such communion in a twofold manner: first, his unlocking towards the Other-being occurs precisely in the element of personal communion in the sense described (now this communion is realized as communion with God in prayer); second, any of the steps of the practice also contain transindividual, intersubjective aspects in which the practice transcends the ascetic's individuality and includes the participation of another individual or of a community; and once again these aspects are constructed as a personal communion. I have pointed out that, according to Foucault, such aspects play a primary role in the ancient practices of the self. They are certainly not less significant in spiritual practice and I will describe them below. But first one cardinally important consequence of this personal nature of the spiritual practice has to be underlined. Personal communion — precisely in the sense described, oriented toward the order of personal being-communion — constitutes a specific type of subjectness, different from the "possessive self" of the ancient practices of the self. This type of subjectness is an authentic "first-person subject," I. *The discovery of the Personality in the ontological and theological plane has as its direct consequence in the anthropological and subject plane the birth of the "first-person subject" in the practices of man.* A human being's ontological unlocking, his attainment of self-openness in being, is based on a particular ever-deepening and flying open communion in which he gradually articulates and reveals himself with extreme fullness; and in this extreme unlocking effort of articulation and revelation, a new subject formation is born. In the ancient practices of the self, the task of the adept, the Pupil, consisted in the acquisition of the Teacher's "true speeches," which is what constituted him as the "possessive self," or "subject," who acquires and receives "his own," "himself," from the Teacher's constituting speech. But in spiritual practice the constituting speech belongs to the adept himself, and this is a personological event: articulating the constituting speech himself, the adept necessarily emerges as a "first-person subject," as an I-subject.

Turning now to the transindividual elements of the spiritual practice, we recall that the most important of them was already mentioned: the practice follows its *organon,* and the *organon* is created and preserved in a certain community, which I have called the *spiritual tradition.* The necessity of the spiritual tradition, the fact that the existence of the spiritual practice is possible only in connection with it and in the midst of it, is one of the most principal characteristics of the spiritual practice. Connected with the necessity of the *organon* of the practice, this characteristic, like the *organon* itself, is rooted, in the final analysis, in the ontological dimension of the practice, the ontological status of its *telos.* Therefore, needless to say that one will not find a spiritual tradition in the ancient practices corresponding to Foucault's "Hellenistic model." But it is possible to notice in these practices certain phenomena and traits that retrospectively can be interpreted as its embryonic or proto-forms. The spiritual tradition is an instrument for the identical translation of authentic experience; and by virtue of the specifics of this experience, spiritual and anthropological, the principal method of translation is living contact, personal communion. *The importance of a continuous chain of personal transmissions of experience* is the constitutive feature of all spiritual traditions; and according to Foucault, Epicureanism came very close to the formation of this feature. Here is what Foucault said: "In the dynasty of Epicurean leaders it is absolutely necessary to go directly back to Epicurus through the transmission of a living example, through personal contact" (421R, 373F). This necessity was motivated precisely by the idea of the identical translation of authentic experience, of "truth": the Epicurean teacher "will speak the truth — that which precisely belongs to the [first] teacher, with whom he has established . . . a mediated connection, though through a series of direct contacts" (421R, 373F). And this personal transmission "will permit the pupil to hear the speech of that first teacher, Epicurus' speech" (421R, 373F). Foucault calls this phenomenon "vertical translation," but we recognize in it nothing other than an embryonic form of spiritual tradition.

Thus, spiritual practice can be realized only in the presence of a spiritual tradition and in the midst of it. It is not just knowledge of the organon that the adept of the practice acquires from the tradition; his application of this knowledge, his implementation of each of the steps of the practice, also includes in certain parts a turning to the tradition, a participation of other adepts of it. Such a transindividual character distinguishes, first of all, the testing of experience, its certification as the true experience of a given stage: an essential element of this testing is dialogic communion with those experienced in the practice. It is easy to see that this communion must be personal communion in the sense described. The acquired experience must be understood and interpreted in the framework of the process of the practice and in terms of its discourse, for which purpose it must be universalized, included in the collective experience of the community of adepts. To achieve this, the experience must be verbalized and articulated with maximum fullness and depth. The articulation is brought to this degree of fullness and depth through communion, which itself becomes ever deeper and thus realizes a kind of "transindividual analogue" of the intentional act, gradually capturing the experience in consideration more and more clearly. This kind of processing of experience is a crucial mission of the spiritual tradition, in the light of which this tradition emerges as a space wholly transparent for (personal) communion: it becomes a *universum of commu-*

nion. Because that which is brought into the space of communion is the ascetic's experience in its full scope with all its triumphs and defeats, with its progress forward, deviations, and falls (and it is only in communion that all these elements of experience are identified as triumphs, defeats, and so on), one can say that the spiritual tradition is also a *space of shared experience.*

These basic definitions of the spiritual tradition make it easy to understand the meaning and function of the concrete forms developed by it. It is clear that, in particular, the institute of spiritual directorship or fatherhood is a natural mechanism of the interpretative, "hermeneutic" sharing of experience, a locus of its communalization, and thereby of its generalization and universalization which the adept needs for further ascent. (It should be noted that the constitutive relation of this institute, personal communion,[10] has nothing in common with relations of power, domination-subordination, and so on; any element of relations of power in personal communion is a distortion alien to its nature.) It is clear too that the practice of *exagoreusis*, the "revelation of thoughts," is one of the transindividual procedures of the hermeneutics of ascetic experience. In its usual form it does not by any means belong to the particularly strict and extreme technologies of the self. *Exagoreusis* included, for example, the question-and-answer sessions (with clarification of individual experiential cases and situations) so widespread among the desert fathers; the practice of these sessions could also be conducted through letters, as for example in the case of the famous elders Barsanuphius and John. A great significance was placed on it in the Studite monasteries, where it took the form of monks' daily visits to the igumen; affirming its necessity, Theodore the Studite explains it clearly and simply in his *Small Catechesis:* "When it is revealed, an inappropriate thought is chased away by the mercy of God, but when it is concealed, it gradually becomes one of the works of darkness. This is the cause of deaths of the soul, of the separation of people from one another, and of self-justifying words engendered by ignorance and error."[11] As is clear from these words, *exagoreusis*, like all the school of Studite monasticism, is not too closely connected with the core of hesychasm, the "Noetic Practice" *(praxis noēta).*[12]

---

10. We do not touch on the question of pastorship, of the relation of the priest to his flock, which is not just based on personal communion but has a very mixed nature in the real social environment.

11. St. Theodore the Studite, "Podvizhnicheskie monaham nastavleniya" (Ascetic Instructions to Monks), in the *Philokalia*, vol. 4 (Moscow: Trinity Lavra of St. Sergius, 1992), p. 631 [Russian translation].

12. "Umnoye delanie." — Trans.

PRACTICES OF THE SELF AND SPIRITUAL PRACTICES

As for Foucault, his attention is concentrated exclusively on the extreme form that *exagoreusis* takes among the "novices" in monasticism (possibly because this is precisely the form described by his sole source in the tradition, Cassian). Already in the earliest period of Christian ascesis, the incipient spiritual tradition produced the anthropological structure that was to become a kind of "atom," or elementary cell for it: the anthropological pair (dyad, binomial) "Disciple-Elder." The disciple is only on the threshold of the spiritual practice, and his "personal communion apropos of experience" inevitably has essential differences. He does not yet have, strictly speaking, the experience of the practice; his problem lies in something else: he has not yet achieved the genuine escape from the world, i.e., the transmutation of his energies and volitions into the alternative, nonworldly order, and he must receive this order from somewhere and assimilate and learn it. Tradition found a successful solution to this problem. Of course, the Pupil comes to some Teacher to learn from him, but he has to learn a very specific thing, not knowledge and not skills, but another and, what is more, alternative order of the self, which the Teacher possesses. Therefore the method of teaching is also specific: it demands that the Disciple's world is incorporated into the Elder's world. All of the Disciple's volitions are eliminated as much as possible (being put in the "grave of the will," as John Climacus says), but the Elder must be experienced and even far-sighted enough to have the Disciple's world fully open up to him (the Elder) and to bear spiritual responsibility for everything that takes place in this world. The "grave of the will" also produces a special, intensified form of *exagoreusis*: in the anthropo-dyad, when *all* (in principle) volitions are excluded, *all* (in principle) intentions, thoughts, movements of the mind, even the smallest ones (as Cassian states directly), must be revealed to the Elder for his permission or prohibition. As experience has shown, it is precisely in this anthropological dyad that out of the Disciple is born the Ascetic whose task is now to perform his own ontological ascent to the Other-being. In other words, the work of the Elder and the essence of what is taking place in the anthropological dyad consist in *ontological*, or *transcending*, maieutics: it is analogous to the work of Socrates, but in a different ontological situation and, as a consequence, in a stranger external form. But, after all, Socrates' actions, too, seemed very strange to the Athenians.

After ancient eldership, it is necessary to mention also the Russian eldership of our time, in which there was developed a new transindividual practice in the hesychast tradition. To be more precise, in Russian eldership the hesychast tradition goes outside its usual boundaries and assumes also a social function. Here the paradigm or dialectic of departure-return (much discussed in recent years) makes its appearance: departure from the world in the spiritual practice can, at the stages of the mature mastery of the practice, be followed by return to the world for spiritual service. This strategy was most widespread in two epochs of intense creative development of the tradition, in the hesychast renaissance in fourteenth-century Byzantium and in the rebirth of heyschasm in nineteenth-century Russia. It leads to the formation of a broad "adhering layer" around a narrow community of adepts of the practice. In this layer the spiritual tradition is influential and serves as a model and reference system for morality and life, thus becoming a significant spiritual and cultural factor in society. The form of "return" that emerged in Russian hesychasm can be described as follows: experienced representatives of the spiritual tradition, the "elders," entered into wide communion with a multitude of simple laypeople of all social strata and classes, serving as their spiritual counselors and teachers. Foucault finds in the Hellenistic practices an analogous strategy of distancing or departure from social activity followed by a return to it. It is instructive to compare the two versions of the same strategy. It is immediately clear that, as a result of ontological otherness of the *telos* of the hesychast practice, the element of "departure" is much more pronounced and radical in it. The "return" is different, too. According to Foucault, in the ancient practices this element is organically included into the practice, completing it: thanks to the preceding "departure," in the "return" we become able to correctly position ourselves in the social reality and correctly organize our connections and activities in it. But for the hesychast elder, in contrast to the Stoic, the return is limited and ambivalent: the elder does not by any means abandon the practice with its alternative order of human energies; entering into the world and finding himself in the thick of people with their worldly order of energies, he does not accept this order but represents what is Other to it, the world and values of the spiritual tradition. Besides this, his work consists not in self-transformation (this work is not the hesychast practice as such but is additional and parallel to it) but in self-giving, in the service of love. And here we come to the most essential feature: the service of the elders represents *a victory of the anthropological over the social*. Although in its external characteristics it is a social activity, their service is actually an entirely anthropological practice, for it is performed exclusively through personal communion and in the element of love, being opposed to all institutional principles, power relations, "matrices of government" (Foucault's formula), and so on. Here it must be taken into account that the elder's contacts are innumerable, usually very brief, and that his authority is enormous; under such conditions one would expect to find formalized or ritualized relations with the presence also of the discourse of power. The phenomenon is thus paradoxical and astonishing, and it also points out to us certain unexhausted resources of anthropological strategies.

With this I have surveyed perhaps all the basic transindividual phenomena in the spiritual practice. But at this point the question should arise: Why did I not examine the phenomenon that for Foucault is the primary one both in Christian ascesis and in the entire Christian culture of the self? That is, why did I not discuss the practices of confession? The answer is simple: *I did not discuss the practices of confession because they do not exist in the spiritual practice.* What? How can they not exist in it??! Here's how. In the transindividual dimension of the spiritual practice there occurs "personal communion apropos of experience" in the universe of Tradition supplemented by certain phenomena that go beyond the frame of this universe (such as Russian eldership). This communion is multifarious; at the initial stage it even features the "grave of the will," but it does not feature (here I am following Foucault's definitions of the practices of confession) "the subject's speech about himself when he is under someone's domination" or "the giving of testimony against oneself." "Personal communion" — in the sense in which we are using it, i.e., oriented toward "personal being-communion," toward interchange of being — and the discourse of power, the power relation, are mutually exclusive; and even in the Disciple-Elder anthropo-dyad the work of ontological maieutics accomplished here cannot be understood in the discourse of power: it is impossible to give birth by command, and the maieutic influence, as in Socrates, is surely not an influence of power.

The exaggerated significance that Foucault ascribes to the practices of confession also reflects (besides the personal factors noted above) the differences between Western and Eastern Christianity. Above I have remarked in passing that the practices of confession are different in these two domains of Christianity, and now I can substantiate this remark. Practices of confession are connected with relations of power, but "personal communion" is not connected with these relations, so that in the spheres formed by personal communion the role of the relations of power, as well as of the practices of confession, is reduced. Moreover, personal communion, being oriented toward "personal being-communion," implies the understanding of personality that corresponds to the patristic "trinitarian personological paradigm," and this paradigm has become firmly established in the Eastern Christian discourse but not in Western Christianity. The conclusion is that, in Orthodox consciousness and in Orthodox religiosity, "personal communion" plays a more significant role than in Western Christianity, whereas the practices of confession play a greater role in Western Christianity than in Orthodoxy. The fate of the *exomologesis* discussed by Foucault can serve as an illustration that confirms this. This rite of public repentance (which does not belong to the sphere of spiritual practice) is very close to the practices of confession but very distant from personal communion. This rite was found to correspond poorly to Christian spirituality, and it fell out of use, but it disappeared a great deal faster in the East, toward the end of the fourth century, whereas in the West it persisted until the seventh century. As for the hesychast practice, it is a phenomenon of Eastern Christianity, and was represented in the West only at a very early stage, by John Cassian, as an exception. Foucault acknowledges that Cassian's ascetic school belongs to the Eastern tradition, but he ignores (at least in the texts known to us) all the consequences of its belonging to this tradition. But these consequences are far from being insignificant, and they include the fact that an adequate interpretation of the transindividual elements of the ascetic practice of the self is provided only by the discourse of personal communion. Evidently, these last remarks are variations on the theme of a general thesis about the "juridical" character of the Western Christian consciousness in contrast to the "personalistic," "existential" character of the Russian and Orthodox Christian consciousness. This affirmation is overly general and clichéd but all the same it is relatively correct.

II.2. From Hesychasm to Synergic Anthropology

The project of the hermeneutics of the subject, as it emerges in the framework of synergic anthropology, is by no means limited to the domain of the "hesychast subject" and of the Eastern Christian discourse which we have been considering until now. As said above, synergic anthropology has as its phenomenal base a particular kind of constitutive anthropological experience: *extreme human experience*, which it describes in terms of "extreme anthropological manifestations." Accordingly, the problems of the hermeneutics of the subject, of the constitution of the personality and the identity of the human being, are formulated and studied here on the basis of the full set of all extreme human experience, and in terms of extreme anthropological manifestations. I call the *Anthropological Border* the set of all extreme anthropological manifestations; this is the key concept of synergic anthropology. The defining characteristic of extreme experience is its connection with *anthropological unlocking*, another basic concept of synergic anthropology already introduced above. Man's unlocking signifies the attainment by him of *openness*, which in synergic anthropology is understood energically: it means the openness of the configuration of the energies of a human being when there occurs the encounter (contact, interaction) of these energies with energies of man's Other. If the unlocking is realized in extreme experience (which is not necessarily so) the Other is not located within the horizon of man's consciousness and experience, and the unlocking is constitutive for the human being as such (see below). As shown above, the key event in hesychast practice is an *ontological* unlocking in which human energies achieve openness to the Other-being, to "personal being-communion" playing the part of the meta-anthropological *telos* of the practice. But this is not the only kind of anthropological unlocking. Beyond the horizon of consciousness there lies, by definition, the unconscious, whose existence affects certain configurations of human energies, so that these configurations correspond to extreme experience, and the anthropological unlocking is realized in them too. The unconscious does not have the status of Other-being, and therefore the unlocking towards it is not an ontological unlocking. In the line of Heidegger, I call it *ontic unlocking*. Furthermore, one should consider that man's unlocking, the openness of the configuration of his energies, is also present in his virtual practices, in the going-out into anthropological virtual reality. The nature of this third, or *virtual,* unlocking is somewhat different: the anthropological openness in it is not an openness towards some representation of man's Other. At this point, however, I will not discuss virtual unlocking in more detail: nor will I try to demonstrate the cardinal fact that *the possible modes of the constitutive unlocking of the human being as such are limited to the three just mentioned.* These questions are outside the problematic of Foucault's project, and my comparison of the two projects does not necessitate their explication.[13]

Thus, synergic anthropology is based on the paradigm of (energic) anthropological unlocking, establishing the existence of three, and only three, representations of this paradigm: ontological, ontic, and virtual unlocking. This conclusion determines the structure of the Anthropological Border. Each form of unlocking is directly connected with a specific form of extreme human experience, with a specific ensemble of extreme anthropological manifestations. But the Anthropological Border is the set of all

13. See note 3, p. xxv.

such manifestations, and hence it follows that it is formed by three domains consisting of anthropological manifestations in which are accomplished, respectively, ontological, ontic, and virtual unlocking. I call these domains, respectively, the Ontological, Ontic, and Virtual topics (topographies) of the Anthropological Border. The Ontological topic includes not only manifestations occurring in hesychast-type practice whose *telos* is personal being-communion. By definition this topic includes manifestations corresponding to any spiritual practices (i.e., to "ontological practices of the self oriented toward a meta-anthropological *telos*"). Spiritual practices are characterized by the presence of an *organon* of their experience. They are created in the womb of the world religions, and the fundamental separation into "religions of Personhood" and "religions of the Cosmos" implies an analogous separation in the ensemble of spiritual practices. I call this separation an ontological bifurcation: the ontological Other to man, emerging as the *telos* of his spiritual practice, can appear either in the personalistic representation (like "personal being-communion" in Christianity) or in the impersonal representation (like the nirvana or Great Void of the Far Eastern practices, qualityless being, indistinguishable from nonbeing). Spiritual practices that are oriented toward the impersonal *telos* also have a stepwise structure, but their steps, or anthropological energic forms, are radically different. Needless to say, progress toward the *telos* here does not occur in the paradigm of personal communion, and it has the character of the gradual self-depletion and deconstruction of personal structures, of "self-leveling." Thus the Ontological topic of the Border consists of two parts, corresponding to the two branches of the ontological bifurcation.

The unlocking in extreme anthropological manifestations provides constitutive anthropological experience, and because of this any specific form of such experience realizes a specific type of constitution of human personality and identity. One can make this proposition even stronger: *only* extreme experience constitutes man as such. A human being can, in principle, choose any of his nonextreme experience to be constitutive for him. In this case he will form up his constitution in this experience, but it will be only a certain *partializing constitution*, which will define him by certain partial, not universally human predicates (such as cultivating a certain lifestyle or belonging to a certain group, party, subculture, and so on). As a result, he will not be endowed with the constitution of the human being as such. Leaving aside such partializing modes of constitution, I shall show that, having described the structure of the Anthropological Border, we can use it to reconstruct the full repertoire of personality and identity structures that a human being can have — that is, to develop a complete hermeneutics of the subject. Before beginning a description of these structures, I would like to underscore that already the fact of the existence of multiple types of the constitution of Man as such is an important and radically nonclassical feature of synergic anthropology. The classical essentialist conception of man dictates that the constitution of Man is determined by his essence, and insofar as essence of Man as such is unique, to Man as such corresponds a unique and universal type of constitution, of personality and identity: namely the classical subject, endowed with essence and substance, and classical substantial identity. In contrast to this, from the perspective of synergic anthropology, every type of extreme human experience (or, what is the same thing, mode of unlocking) defines a certain type of constitution of man — precisely of nonpartialized man, of Man as such. In other words, it defines a certain complete and full-fledged representation of Man as such. Man is pluralistic in his nature: he is in fact an ensemble of self-sufficient, radically different beings or creatures corresponding to different modes of anthropological unlocking, or to different topics of the Anthropological Border. Let me examine this ensemble and its characters.

First of all, there are three main types, corresponding to the three basic mechanisms of unlocking: *Ontological Man — Ontic Man — Virtual Man*. For each of them there is a principal sphere of his realization: spiritual practices — patterns of the unconscious — virtual practices, respectively. There are also three additional "hybrid" types. The three basic types of extreme anthropological manifestations forming the Anthropological Border can coexist, for manifestations whose extreme nature is implemented simultaneously by two (or all three) basic mechanisms are possible a priori (and can actually be identified a posteriori). This means that the three above-described topics of the Anthropological Border overlap, and their overlapping also forms three domains, which we call "hybrid topics." The hybrid structures of personality and identity corresponding to them are very characteristic of and symptomatic for present-day anthropological processes, but for my present theme it is sufficient to consider the basic types.

Ontological Man is constituted by energies of the Other-being. This means that his identity is formed in the process of ascent on the ladder of energic forms, a ladder oriented toward the Other-being as the meta-anthropological *telos* of the whole process. The identity constituted in this way can be called *participative identity*, for it is attained by means of participation in the energies of the Other-being. The Other-being emerges here as the supreme source of the identity: it is the sole holder and guarantee for the principles producing and grounding identity, the sole entity that has the ability and power to constitute out of itself and delegate identity. Thus, it is also the sole entity that is endowed with absolute identity or, more precisely, with identity in the exact and complete sense; and all the structures and types of identity realized by empirical (quasi-)persons in man's mode of being must be regarded as only partial and imperfect (participative!) actualizations of the true identity belonging to the Other-being.

The further properties of Ontological Man depend on the ontological bifurcation described above. The personal and impersonal representations of the ontological Other, God as Personality (personal being-communion) and the impersonal Absolute (nirvana or the Great Void, the principle beyond the opposition of being and nonbeing), serving as the *telos* of anthropological strategies, constitute radically different types of human personality and identity (even though these different types have the same participative nature, i.e., they are formed by the same generative principle — by the principle of energic participation in the ontological Other). In their purest form, these types are realized in corresponding spiritual practices, such as in hesychasm in the personalistic paradigm and the Tibetan Tantra or Taoism in the impersonal paradigm. The concrete character of these types affords striking evidence of the formative role of the meta-anthropological *telos*.

We have already cursorily described the structures of personality and identity formed in the ascent to the personal *telos*. This *telos* is personal being-communion, or the perfect Personality, or the Trinity of consubstantial Persons-Hypostases whose mode of being is conveyed by the concept of *perichoresis* and by the fundamental identity: *perichoresis* — love — (personal) communion. It is clear that this identity also defines a certain type of self-identity, which can naturally be called *trinitarian self-identity*. As I have already pointed out, this is ideal, absolute identity, which is realized fully only by (personally represented) Other-being. As for empirical man in the horizon of being-there, he realizes imperfect participative identity, which presupposes energic participation in the trinitarian identity and is formed in the ascending spiritual-anthropological process. As said above, this process takes place in the element of "personal communion," that is, such communion which, becoming ever deeper, approaches interchange of being, *perichoresis*. Here this communion is of a double kind: in the encounter with the energies of the *telos* and in the transindividual, intersubjective technologies that enter into the *organon* of the spiritual practice. In this communion there takes place a building-up of personality, a generation of more highly organized anthropological energic forms possessing a higher degree of connectedness (which, in particular, signifies an increase in the clarity of self-consciousness). This is facilitated by the bidirectional principle "expulsion of images — warming of feelings" (Theophan the Recluse's formula) by virtue of which the process acquires an intense existential and emotional character and corresponds to the "warming-up" of an individual's inner reality. In sum, the participation in trinitarian, "perichoretic" identity which takes place with the substantial help of emotional activities leads to strengthening, articulation and growth, and enrichment of the reduced and defective structures of identity corresponding to ordinary existence.

Almost diametrically opposite to this are the structures of identity in the impersonal paradigm. The impersonal *telos* is the Absolute, which is deprived of all dynamics and all structures. As a result, the steps of the ladder ascending (or descending, if you like) to this Absolute represent energic forms that are less and less complexly organized. The anthropological energic structures are successively self-disassembled by an individual; they are taken apart, blurred, and dissolved. In Neoplatonic mysticism, which also belongs to the impersonal paradigm, this process acquired the name *haplosis*, simplification. The methods of these processes are opposite to St. Theophan's rule: they realize "contemplation of images — expulsion of feelings" since they are based chiefly on various techniques of meditation and gradually lead to the cessation of all emotional activities. Instead of a "warming-up" of the inner reality which accompanies the building of the personality, leading to the dynamically structured trihypostatic being-communion — instead of this we have a "cooling-down" of the inner reality and a (self-)destructuring, or self-dismantling. And it is clear that, at the end of this process, personal identity and self-consciousness dissolve and disappear completely. As one of the Yoga Sutras of Patanjali formulates it, as the *telos* (Samadhi) is approached, "self-consciousness is deprived of every proper form of itself and is wholly dissolved in the essence of that which is contemplated, where Being and Nonbeing are indistinguishable" (Sutra III, 3).

Ontic Man. The ontological bifurcation divides the domain of Ontological Man into two parts which are very different from each other. But however great their differences may be, the differences between the different basic topics of the Anthropological Border are far greater. In the broad sense one can say that the most general difference between the Ontological and Ontic topics consists in the *replacement of ontology by topology*. The action of the unconscious brings forth an anthropological process with a dynamics radically different from the ontological dynamics, which produces the generation of an ascending hierarchy of anthropological energic forms. In this case the anthropological dynamics is constituted by the Source-beyond-there which belongs not to the other but to the same ontological horizon, to the "being of the essent," and this dynamics is usually characterized as *topological*. The term originates in the fact that the presence of an energy or power source in the same horizon of being but beyond the horizon of consciousness and experience acts as a topological anomaly: due to the influence of this invisible and extra-experiential source, the topology of the world of consciousness and experience stops being the habitual topology of the Euclidean world and becomes one of the "nonclassical" topologies (even though the term "nonclassical" is not employed with reference to topologies) — curved, multiply connected, discrete, etc. This general feature is reflected, in particular, in the fact that typical phenomena conditioned by the unconscious are called "patterns" or "figures" ("figures of the unconscious" is one of Jung's terms), which points to their topological nature. The richest base for identifying and investigating the topological nature of phenomena of individual and collective being is provided by Deleuze's philosophy, which enunciates a "topological thinking" and develops an integral philosophical-topological discourse with an extensive arsenal of concepts of a kind unusual for philosophy (folds, joints, bends, tears, turnings inside out, underlayers, etc.). But in exploiting this arsenal, one must take into account the fact that Deleuze's own attitude toward topological thinking is one of uncritical enthusiasm. He is inclined to absolutize the topological vision of reality and often ignores the fact that there are many spheres of phenomena and processes of a different nature for which this vision is inadequate.

Topological dynamics generate extremely specific structures of identity, and one can say that the identification and correction of these structures constitutes the principal content and goal of all the refined methods of psychoanalysis, if not of this discipline as a whole. The fact that these structures are viewed as requiring correction, therapy, attests that they are close to being pathological formations, if not actually being such. This already reveals a profound difference between the topics of the Anthropological Border: with respect to human identity the Ontological and Ontic topics as well as the representations of the Other corresponding to them, turn out to be diametrically opposite to each other. If the Other-being is the source of identity, constituting and filling it, then the unconscious, a topological anomaly in the world of consciousness, is a source of influences that deform and traumatize identity. The key characteristic of the whole Ontic topic as such is a *disconnected topography of consciousness*. As a result of this disrupted connectedness of the world of consciousness there is generated a broad spectrum of various "subnormal phenomena" — of disorders and dysfunctions at the lower levels of consciousness, such as various instances of disharmony and incoherence of behavior and action, blunders, "Freudian slips," etc. It is clear that all these phenomena express defects of identity, and such topological defects and disruptions should be viewed as specific characteristics of identity structures in the Ontic topic.

Virtual man is a relative newcomer to modern reality, and his nature and properties have as yet been insufficiently understood. Nevertheless, the principal features of his identity structures are immediately evident. The key characteristic of virtual phenomena as such is underactualization: every such phenomenon is an underactualized copy of some actual phenomenon, i.e., a copy approximately coinciding with the latter, but with some basic predicates subtracted. This general characteristic is also applicable to identity: one can say that *Virtual Man is endowed with underactualized identity*: his identity is incomplete; it is deprived of some of the basic elements or structures of full-fledged human identity. Because of this characteristic of incompleteness or deprivation, the virtual type of identity resembles the ontic type, which we also characterized as having certain defects of identity. On the other hand, it also resembles identity structures in spiritual practices with an impersonal *telos*, where, as we have said, a blurring and dissolving of these structures takes place. Both comparisons are useful; I will examine them in more detail in order to better elucidate properties of virtual identity.

In spiritual practices, ascent to an impersonal *telos* presupposes a conscious, intentional, and controlled dissolution of an identity that has been fully developed and articulated at the initial stages. By contrast, in virtual practices identity is initially underdeveloped and underactualized, and by no means is it subjected to some intentional and controlled change (it is the capacities for self-control that are most underactualized and defective in Virtual Man). As for patterns of the unconscious, each form of them represents a certain topological phenomenon or effect implying a specific defect (a deformation, pseudomorphosis, truncation, or some other disruption) of identity, so that the defects arising here are far from arbitrary but, on the contrary, can be systematized, classified, and reduced to types that serve as symptoms of the corresponding patterns — neuroses, phobias, etc. But, generally speaking, in the underactualized identity of Virtual Man any elements or structures of identity can turn out to be not actualized. In other words, absolutely arbitrary defects of identity are possible here, and this

Spiritual Practice, Synergic Anthropology, and Foucault's Project

clearly means that, compared with the Ontic topic, in the Virtual topic the destructive and decomposing tendencies in the sphere of human identity take on a more profound and radical character.

II.3. Historical Sequence of Anthropological Formations

It is indisputable that the anthropological experience of the present time corresponds to some sort of radically new, maximally plastic, changing, and polyphonic image of Man — of Man who is capable of choosing stunningly different scenarios of self-realization. Synergic anthropology proposes a specific way and method for the reconstruction of the whole spectrum of these scenarios. It is based on the key observation that the ancient paradigm of synergy, that is, of anthropological unlocking, can be newly conceived and generalized — so as to become, after having incorporated the new modes of unlocking, a universal dynamic paradigm of human constitution. This way is not connected with the postulates and limitations of classical essentialist anthropology, and thanks to this, it leads directly to the conclusion of principle: the fundamental structure of human existence is such that *man is pluralistic,* not only at the obvious level of the possibility of different scenarios of self-realization, but also on a much deeper level. Even at the level of the basic structures of his constitution, Man is already, if you will, a community. It would be more correct to say that, instead of a single "human being" (which is how philosophy has always conceived Man), there exists an "anthropological space" in which is actualized a collection of beings having fundamentally different constitutions of personality and identity and unceasingly undergoing transformations into each other of all kinds. Ontological Man, who realizes the paradigm of Christocentric deification or that of the Kalachakra Tantra, is one of these beings, but he is not alone in the anthropological space. Topological Man, who realizes patterns of the unconscious, is one of these beings, but he is not alone in the anthropological space. And so on.

It is instructive to compare these conclusions with the positions of Foucault's project. The basic part of this project, presented in the 1982 Course, reconstructs three models or "great formations" of the practice of the self, which are almost wholly limited to the millennium from the fourth to fifth centuries BCE to the fourth to fifth centuries CE. Needless to say, all three formations embody a unitary, not a pluralistic vision of Man's nature since there was no alternative to the unitary vision in that epoch; meanwhile, Foucault's own views on Man as such remain somewhat implicit here. We can only observe that Foucault's predisposition (which is obvious also in this Course) to assert a purely immanent constitution for man conceals a hidden tendency to favor the pluralistic anthropology. But this tendency is expressed much more explicitly in his less major texts (interviews and articles) of the late period. As I have shown, in his project of the "esthetics of existence," Foucault puts forward a "neotribalist model" that represents the human community as a combination of "subcultures," or minorities, defining themselves according to methods for obtaining pleasure. This model can be viewed as an extrapolation of the "Hellenistic model," but a radical extrapolation that takes the immanence of man's constitution to the maximum, to an "immanence most fierce," to use the joking expression from the slang of Russian intelligentsia of very old times. If the method used for obtaining pleasure is taken as the principle of human constitution, this, needless to say, is a profoundly immanent constitution. Not based on extreme experience, which is the only kind of experience that can constitute man as such (see above), this immanent constitution partializes man. But here it must be said that, in this paradigm of constitution, the immanence is so complete that the boundary and difference between partial (or partialized) and nonpartial man disappear. Man in each of the minorities has the right to suppose himself to be "man as such," and not "partialized man," insofar as it has become impossible to answer the question: *The partialization of what (of whom) is the man of this minority?* All minorities have an equal anthropological status; and the "majority" in this model does not exist at all; it is excluded from the discourse. "There is no universal; there is only the singular," declares Deleuze, who completely shares Foucault's views in these matters. As for the property of pluralism, we see that the neotribalist model is indisputably and radically pluralistic: every subculture is self-sufficient, a complete realization of Man. But this pluralistic character is of a different nature than in synergic anthropology. The latter does nothing more than affirm that Man has initially a certain inner pluralistic character, implanted in the very structure of his constitutive, extreme experience. This affirmation does not by any means imply that an individual must embody and cultivate this pluralistic character in his strategies (just as, for example, the affirmation that men can be liars, thieves, and swindlers is not tantamount to asserting the desirability or inevitability of such behavior). Making this affirmation, synergic anthropology does not yet decide what an individual's attitude should be toward this trait of his (it sees here a special problem, requiring a very different context for its discussion). In contrast, Foucault loudly advocates the "neotribalist model" as an anthropological strategy; he presents it as his own project of an anthropological future. And now it becomes somewhat more clear how closely is this project linked with the projects oriented toward the Post-human.

It should be noted that, in his book on Foucault, Deleuze's cursory reading of Foucault's anthropological project is quite different from my reading of it as a "neotribalist model." As Deleuze finds, Foucault's conception, when applied to Modern Times and the contemporary world, is reducible, as before, to three great anthropological formations, but these formations do not coincide with the three identified by Foucault himself when he studied "the millennium of the care of the self." All these formations are generated by means of the same universal dynamic paradigm, which consists in the encounter and interaction of man's own inner forces with certain external forces: "[T]he forces in man constitute a form only when they enter into a relation with external forces."[14] The three new formations, which, too, successively give way to each other, correspond to different representations of the external forces; according to Deleuze, they are as follows: the "form God," constituted by "the forces of elevation to the infinite" (the seventeenth and eighteenth centuries); the "form Man," constituted by "the forces of finiteness," in the capacity of which we have Life, Work, and Language (from the nineteenth century); and finally the "form Superman," which is still in an embryonic stage and which presumably will be constituted by the forces of the "finite-unlimited" *(fini-illimité)*, forces that are present, for example, in computer chips, in elements of the genetic code, and in agrammatical structures *(des agrammaticaux)*, which "take their revenge on the signifier" in texts of modernistic literature. This last formation is directly traced back to Nietzsche's idea of the Superman; according to Deleuze, this is the only adequate embodiment of this idea in the present time.

Several remarks are needed in connection with this reading of Foucault. First of all, Deleuze's dynamic paradigm clearly coincides with our paradigm of anthropological unlocking, and by the same token, in Deleuze's reading, Foucault's method of constituting anthropological formations coincides almost entirely with the analogous method in synergic anthropology.[15] But I find, however, that Deleuze's description does not fully corre-

---

14. Gilles Deleuze, *Foucault* (Paris: Les éditions de Minuit, 1986), p. 139.

15. Still they do not fully coincide. Deleuze speaks of forces, not of energies, and his discussion follows in the discourse of topological dynamics. In this discourse ontological unlocking is excluded ("the forces of elevation to the infinite" are far from being the same thing

spond to Foucault's discourse and that the paradigm of unlocking is more characteristic of Deleuze than of Foucault. In particular, Deleuze's thought, much more so than Foucault's, is concentrated on the dynamics of reality, on the dynamic structure of phenomena and forms, whereas Foucault's thought is more concerned with their integration into the historical and cultural context. In describing the constitution of anthropological formations, Foucault does not care much about the in-depth analysis of the dynamic paradigm of this constitution, but he does strongly underscore the immanent character of the latter (as we recall, this is a sensitive topic for him). But the paradigm of unlocking presents an expressly nonimmanent method of constitution, and therefore Deleuze's treatment does not reflect the pathos of immanence, which constitutes such a characteristic feature, if not the leitmotif, of Foucault's last project. It is necessary to clarify, however, that Deleuze's schema of three formations refers primarily to the "early Foucault"; its description sends us back almost exclusively to *Les mots et les choses*. As for the stage of the practices of the self, Deleuze, as already said above,[16] insistently (and quite correctly) indicates that in this stage Foucault does not by any means bring the concept of the subject back into his philosophy and that his "history of the subject" is more precisely a history of forms of subjectivation represented as a "production of modes of existence," so that here anthropological reality is described as a set of such modes. One can naturally associate these modes of existence, which Deleuze also calls "styles of life," with the "subcultures" in Foucault's late interviews on the basis of which his last project can be characterized as a "neotribalist model." Deleuze, however, conceives their principle of constitution as broader than just a "method for obtaining pleasure" (in particular, he says that the modes being produced "include even suicide"). In sum, Deleuze's interpretation of Foucault's late project agrees fairly well with my own, except that his is somewhat broader and in particular he appropriately reminds us that the subculture of suicide should not be excluded from the set of subcultures. There remain only two points of divergence. The first is the failure to take into account the strong accent on immanence of all Foucault's constructions of subjectivation; the second consists in the fact that we absolutely do not see in Foucault's project any elements of the "form Superman." The sub-

---

as "the energies of the Source-beyond-there"); in addition, the categories of internal and external (with reference to man) are understood differently than in synergic anthropology.

16. See our discussion above, in the section "The Language of Foucault's New Conception of the Subject."

Spiritual Practice, Synergic Anthropology, and Foucault's Project

cultures considered by Foucault — those of the gays, the sadomasochists, etc. — are far from being so futuristic and post-human in character, and Foucault does not by any means connect them with Nietzsche's Superman (the discourse of the Superman is completely absent in Foucault, and here Deleuze ascribes to his friend his own predilection for this idea).

As we have seen, in Foucault's project — both in its principal part, devoted to the "millennium of the care of the self," and in sketches devoted to the modern "history of the subject" — an important role is played by the ordering of anthropological formations in diachrony, in historical sequence. Synergic anthropology, too, carries out such a "historical ordering" of the formations introduced in it, establishing their historical sequence and tracing the causes and mechanisms of their succession. My starting point in this is the fact that, in any period of history, specific anthropological formations predominate; and these formations can in principle be found in the fund of all basic formations connected with the Anthropological Border. However, historical survey discovers also situations when the adequate formation cannot be found in this fund. It will be shown that these situations represent special cases that do not contradict the principles of synergic anthropology. On the contrary, this survey confirms that the constitutive anthropological experience is the extreme experience, i.e., the experience of unlocking, which has three (and only three) principal forms.

In the Anthropological Border there are only three fundamental formations: Ontological Man, Ontic (or Topological) Man, and Virtual Man. But the early epochs of history already present to us an example when the dominant anthropological formation was not one of these three. Knowing that in these epochs the dominant anthropological relation was the man-God relation, one would apparently be right to conclude that the dominant anthropological formation for these epochs was Ontological Man. But that would be a hasty conclusion. The unlocking paradigm in its three "pure forms," which are distinctly different from one another, was not present from the very start; primordially, a long period of time was needed for its forming-up. During this period anthropological unlocking was actualized in fused or amalgamated proto-forms, and although man was oriented towards the Man-God (or "divine powers") relation as his formative relation, he was not Ontological, but "Pre-Ontological," Man. It took him a long time to develop the paradigm of ontological unlocking. Man had to accomplish a colossal labor of reason in order to achieve self-identification in being and separate the ontological from the ontic; and one more significant stage of the labor of reason and of religious consciousness was needed before there could emerge a Christian ontology of splitted being and the ontological unlocking toward the Other-being. These labors can naturally be associated with the epoch and the process that Jaspers called the axial age and the formation of axial civilizations. One can say that synergic anthropology discloses anthropological contents of these concepts, thus complementing them with the anthropological dimension.

The initial anthropological formation was defined by archaic forms of religiosity in which man's extreme experience was represented only by obscure, unarticulated impulses and strivings of a mixed nature: man had not yet separated the manifestations in which his self-identification in being was realized from those in which his relations with the unconscious were realized. There was still no distinction between the ontological and the ontic in his consciousness.[17] One of the principal primitive religious forms of this kind was shamanism, which Foucault refers to more than once as an ensemble of practices preceding the practices of the self of ancient Greece and influencing them: "Exercises that apparently could be traced back to shamanism ... were at the origin of the spiritual exercises which were developed in Greek

---

17. It is useful to make a clarification here because such mixed-nature phenomena involving the superposition of forms of extreme experience are present in the case of modern man too. Both in spiritual practices and in modern psychology and psychiatry we often encounter phenomena in which the ontological (the authentically religious) is mixed and fused with the unconscious. In synergic anthropology the "hybrid" topics of the Border correspond to these phenomena of the superposition of forms; in particular, the "topic of spiritual delusion (*plani*, in Greek)," including phenomena of ascetic demonology, corresponds to the mixing of elements of spiritual practice with patterns of the unconscious. But it is important to note that, generally speaking, these phenomena of mixing in the modern consciousness are not identical to the original archaic fusion or better indistinguishableness of the topics of the Border. In the contemporary consciousness there occurs the real mixing and/or fusion of two different elements the distinction between which is known to and recognized by the consciousness. By contrast, in the archaic consciousness two different elements do not yet exist, and what we have properly speaking is not mixing but an absence of separation, ignorance of the fact that there are (albeit potentially) two elements, not one. These two situations must generate different types of processes in the consciousness. Insofar as this distinction in the types of processes has gone virtually unnoticed in scientific studies, there arises the danger of mixing also in their sphere: the mixing of shamanism with phenomena of ascetic demonology and with spiritual practice. Synergic anthropology makes it possible to clearly distinguish between all three phenomena, linking them with different anthropological formations: the first is linked to Pre-Ontological Man, the second to the hybrid "topic of spiritual delusion," the third to the Ontological topic. Such a distinction should be useful in light of the heightened interest in shamanism and in other types of primitive religiosity stirred up by present-day psychological technologies and spiritual movements.

philosophy" (451R; 399F). From the point of view of synergic anthropology, shamanism is the purest example of a proto-religious formation corresponding to the initial inseparableness of the Ontological and Ontic topics: here the set of *goals* of spiritual practice organized by the urge of transcending is combined with a set of *means* wholly based on the energy of the unconscious. The further evolution of ancient religiosity demonstrates clearly the process of the separation of the topics: one can interpret the ancient theogony, the war between and the separation of the Olympian and chthonic gods, as a mythological reflection of the anthropological process of the formation of consciousness. In this process there occurred a separation between efforts at self-determination in being and patterns of the unconscious (between the "upper" and "lower" topics of consciousness). But neither shamanism nor Olympian mythology nor even mature Greek and Greco-Roman thought succeed in going-out of the framework of Pre-ontological anthropological formation. There is as yet "no subject" here, Foucault says. "There is no subject because there is no Personality and no ontological unlocking," synergic anthropology adds in agreement. But at the same time, Andrey Rossius, a specialist in the study of antiquity, is also profoundly right when he reminds us that "the Greeks possessed . . . extremely clearly etched forms of tragic subjectivity."[18] And, expectedly and inevitably, in the Hellenistic model of the practices of the self as it is represented by Foucault, one finds many aspects in which the Pre-topical formation comes very close to Ontological Man. The "possessive self," transforming itself, acquires elements of the fundamental structure of an authentic "first person." But a dividing line between them always remains: in this I am in full agreement with Foucault.

Incipit Persona (although, as has been often explained, the Latin *Persona* is a very inadequate way to convey the concept of Personality, which was formed in a Greek-speaking environment and came to occupy a fully adequate place only in the Eastern Christian discourse). A brief description of Ontological Man was given above. The purest form of his self-realization is spiritual practice existing in unity with spiritual tradition and following, in conformity with ontological bifurcation, one of two paradigms: personal

---

18. Andrey A. Rossius, "Vvedenie" (Introduction) in Friedrich Nietzsche, *Rozhdenie tragedii* (The Birth of Tragedy) (Moscow: Ad Marginem, 2001), p. 37 [Russian translation]. With this reminder is associated a whole set of problems of pre-ontological and pre-topical anthropology and personology. I only point out that the analysis of "tragic subjectivity" leads to a general question not only anthropological but also culturo-philosophical in nature: How is the separation of the Ontological and Ontic topics related to the famous dichotomy between the Apollonian and Dionysian principles in the consciousness and culture of antiquity?

or impersonal. In setting up the historical sequence, I will not include the world of Far Eastern impersonal practices. As for practices of the personalistic type, they were developed in Eastern Christianity, and the main epoch of their formation and cultivation lasted (as in the case of the ancient practices) a millennium. If, according to Foucault, "the epoch of the care of the self" stretched over the millennium from the fourth to fifth centuries BCE to the fourth to fifth centuries CE, then "the epoch of the hesychast organon" stretched from the fourth to the fourteenth centuries CE, culminating in the hesychast renaissance in Byzantium. There takes place here the "discovery of the Personality," and the Personality constitutes the foundations of a new anthropological formation. A cardinally new position of consciousness arises, the position of "appearing in person before Person,"[19] which a teacher of hesychasm conveys as follows: "Take care to establish yourself in the conviction that, within you, you always have a Person who is looking at you and seeing everything that there is in you."[20] And in the framework of this position, which has become a constituting and organizing principle of man's inner world, is born, finally, the authentic "first person," I, called the "subject," the "person," the "personality" and by other names. Recalling the characterization of the ancient type of subjectness as an "ensemble of impersonal forces," one can say that the event that has taken place corresponds, mutatis mutandis, to Freud's famous formula which defines the goal of the psychoanalysis of the subject: *Wo Es war, soll Ich werden*[21] (from which follows that the work of Christ can, if you will, be considered as including the work of the psychoanalyst). To the "first person" corresponds "personal communion" as a special form of human communication progressing into the depths from the exchange of information to the exchange of being according to the model of *perichoresis*.

However, as a result of secularization processes, beginning with the Renaissance, man gradually comes to reject ontological unlocking as a principle of his constitution. The dominant anthropological significance of the relation to the Other-being is disavowed, and this relation is more and more moved to the margins and pushed aside. But the unconscious does

---

19. The Russian *"litsom k Litsu"* means literally "face to Face." — Trans.

20. Theophan the Recluse, "Umnoe delanie o molitve iisusovoj" (Noetic Practice Related to the Prayer of Jesus) in *Sbornik pouchenii Sv. Otcov i opytnyh ee delatelej* (Collection of Teachings of the Holy Fathers and of Experienced Practitioners of the Prayer of Jesus), ed. Igumen of the Valaam Monastery Khariton, 3rd ed. (Sergiev Posad: Svyato-Troitskaya Sergieva Lavra, 1992), pp. 132-33 [Russian translation].

21. "Where the id was, the ego shall be." — Trans.

Spiritual Practice, Synergic Anthropology, and Foucault's Project

not take the place of the Other-being: the new anthropological formation, once again, does not belong to the topics of the Border. The centuries of the Renaissance, as well as of the Modern period, are an epoch when the unconscious has not yet become an object for reason, the relations with it are not reflected on, and it is absent from the horizon of consciousness. Thus, man's relation to the Ontic border has not yet been formed while his relation to the Ontological border is being actively pushed away, and this means that Man is losing his relation to the Anthropological Border as such. His strategies and practices no longer reflect this Border's presence and as a result he stops seeing it and the very idea of its presence is effaced in him. It is very natural that the idea of the infinite would occupy the center of consciousness during such an epoch. Man creates the conception of an infinite universe and strives to understand himself as a subject of knowledge who is constituted from his cognitive-instrumental relation to this universe. If the universe is infinite, it follows that the process of its cognitive-instrumental exploration and mastering is potentially infinite too. *Ergo*, Man, who is constituted in this process, is himself infinite. *The conviction is formed in man that he does not have any borders.* We can find a great many direct expressions of this position in the culture of the Renaissance and Enlightenment: for example, in a well-known treatise by Condorcet (who at the time was hiding from the Jacobin terror and soon committed suicide in prison) we read the following: "There was no predetermined limit to the development of human abilities . . . man's ability to perfect himself is truly borderless, the only limit to successes in this perfection is the duration of the existence of our planet."[22] One can say that what we have here is an intermediate anthropological formation that can be called the formation of *Borderless Man*. Up to nowadays this formation is still very influential and widespread even if it is no longer dominant. In many ways Western man continues to accept the ideals and reference systems of Borderless Man, and it is precisely this formation that is most closely linked to technological progress and technogenic civilization. It is thus worthwhile describing it in more detail.

The defining elements of this formation are the idea of the infinite universe and of the infinite process of knowledge, as if rectilinearly moving farther and farther ahead. It is clear that the foundations of this formation also necessarily include the ideas of infinite rectilinear motion and

---

22. Marie-Jean Condorcet, *Eskiz istoricheskoj kartiny progressa chelovecheskogo razuma* (Essay on the Historical Picture of the Progress of Human Reason) (Moscow: Sotsekgiz, 1936), p. 15 [Russian translation].

of a universe ordered by a rectilinear reference system. Borderless Man is Descartes' subject of knowledge, living in Cartesian coordinates. He resembles also Deleuze's "form God," which is constituted by the interaction of human forces with external forces represented as "forces of elevation to the infinite." Further, the methodology of synergic anthropology requires a reconstruction of the paradigm of man's constitution realized in this formation. The constitutive relation chosen for himself by Borderless Man is "the cognitive-instrumental relation to the infinite Universum." There is no doubt that man unlocks himself in this relation, and we have to identify the type or mode of this unlocking. As is clear first of all, the latter does not coincide with any of the types of unlocking defining the basic topics of the Anthropological Border. All the other types are either mixed (hybrid) or partializing, i.e., representing unlocking toward some particular essent. The unlocking structure of Borderless Man turns out to be quite interesting.

One can see, first of all, that the relation to the infinite Universum is essentially deficient as a constitutive relation. The defining characteristic of this Universum, infiniteness, is by its nature a negative principle, the assertion that an object lacks definite boundaries. In such an object it is, figuratively speaking, impossible to find any solid support. Speaking more precisely, the Universum, if it is defined by the predicate of infiniteness, cannot be the full-fledged Other for man — it cannot be an entity in the encounter with which an individual experiences its formative and constitutive effect. In defining himself by means of the relation to it, an individual becomes only endowed with the "possibility of infinite development," and this predicate (in contrast to the predicates of the relation to God and the relation to the unconscious) does not yet lead to any well-defined paradigm of human constitution. This means that, in relation to the infinite Universum, only an illusory unlocking is realized, an unlocking that does not produce a full-fledged principle and mode of anthropological constitution. An individual who defines himself by this relation cannot obtain from it any finished constitution; at best he can only obtain some elements of a constitution. Thus, Borderless Man is underformed and underdefined in his constitution, and in order to make this constitution complete, he borrows or "forages" materials for his personalistic structures from various sources. Depending on the epoch and on the circumstances, he can assimilate elements of the Ontological and Ontic topics, different partializing types of unlocking. Insofar as his defining relation includes an instrumental or technological component, Borderless Man necessarily performs unlocking in instrumental-technological activity and obtains in it elements of the constitution of *Homo technicus*. This is one of the basic forms of the partializing constitution, and Borderless Man, being in close proximity to it, is easily reduced to this form. In sum, the described properties characterize Borderless Man as a mixed and intermediate formation, deprived of its own independent paradigm of anthropological constitution.

It is also clear that this formation is wholly based on an absolutization of reason, on a kind of cult of reason. It is precisely reason that actualizes man's relation with the universe. And if this relation is to be constitutive for man and possess all the necessary properties, man's reason must be constitutive as well as infinite and all-powerful, at least potentially. This conception of reason is inseparable from Borderless Man. It implies that in Borderless Man reflection takes place only upon those boundaries or borders of reason that reason itself posits out of itself (this reflection was masterfully accomplished by German idealism). The postulate of infiniteness denies the existence of the Anthropological Border and, in particular, it excludes the reflection on the Ontic topic and man's relation to the unconscious. This must be compared with the situation of the Ontological Man. The latter did not by any means ignore the unconscious. In his epoch reflection was not developed on the level of classical metaphysics; relations with the unconscious were not explicated in scientific or philosophical discourse, but they were far from ignored. Ontological Man was elaborating his relations with the unconscious: calling its manifestations passions, he recognized their important role and constructed practices that impeded them. As for the Borderless Man, in the process of excluding the man-God relation, he gradually discarded all the basic structures of this relation, including the practices of struggle against the passions (meanwhile he most radically reinterpreted the very concept of passion and diametrically transvaluated it). As a result, if Ontological Man possessed a definite constructive relation to the unconscious, then Borderless Man was only capable of denying the unconscious. But in this situation the effects of the unconscious could develop and grow without hindrance; the role and place of the unconscious in anthropological reality were gradually becoming more prominent. I will sum up this conclusion by a brief aphorism: *the cult of reason leads to the kingdom of madness.*[23] Foucault's early historical studies that reconstructed the relation of Borderless Man to madness, as well as to practices of punishment and repression, confirm and richly illustrate this thesis.

---

23. There is a play here on the words *razum* (reason) and *bezumie* (literally: unreason or mindlessness). — Trans.

"In our time neurosis has taken the place of the monastery." This saying of Freud's precisely expresses the change in the dominant anthropological formation; it just needs to be added that "neurosis" entered the forefront not directly after the "monastery," but after the "physical laboratory," which prepared and facilitated its entry. Ontic Man (he is also Mad Man and Topological Man) achieves his dominant position at the end of the nineteenth century and keeps this position during nearly the entire twentieth century. At the middle of that century it is not even a Freudian but a major representative of neo-Hegelianism, Jean Hyppolite, who asserts that "the study of madness is at the center of anthropology, at the center of the study of man." A sharp activation of the Ontic topic occurs; Freudianism becomes a worldview and the patterns of the unconscious become the dominant form of extreme human experience. Esthetic, artistic practices played a most considerable role in establishing the dominance of Ontic Man: artistic creativity has always concealed within itself threads of a profound and intimate connection with the domain of the unconscious. One can clearly trace how the culture of modernism moved in its evolution from the sphere of Borderless Man (where it still bordered on romanticism) to the kingdom of Mad Man. Initially, the "experience of madness" was experienced by man the artist as a supremely authentic, responsible, and mortally dangerous experience (as was attested by many striking and tragic fates). But gradually — already in surrealism, for example — the artist begins to consciously exploit this experience.

The next step after conscious exploitation is performance-staging, and this is already a strategy of underactualized manifestations which belongs to the Virtual topic. The anthropological dynamics accelerates, and with the approach of the third millennium Virtual Man comes to the forefront. We will not judge to what extent the dominance of this anthropological formation is already established now. We will only observe that the epoch of the dominance of Virtual Man is necessarily acquiring one defining quality: the very definition of virtual phenomena excludes the possibility of the creation of new forms. In a certain sense this whole epoch is only a great remake of all the other anthropological formations with their epochs.

Behind the above-described evolution of Man one can glimpse a certain general direction, a unifying dynamics, and this is a dynamics of successive decline, of an attenuation of the form-building, creative energy of the human being. One can call this global meta- or mega-process "Man's downward slide on the Anthropological Border." The discussion above makes visible this dynamics of decline, in particular, with respect to hu-

Spiritual Practice, Synergic Anthropology, and Foucault's Project

man identity: in describing Ontic and Virtual Man we specially stressed the defective character of their identity structures, and we also pointed out that this defectiveness is exacerbated with the transition from the Ontic to the Virtual topic. But we must also add that the possibilities of the virtual remake are by no means limitless. A more and more complete immersion in the Virtual topic brings forth ever-increasing risks of the loss of control of ourselves and of the systems of the support and security of our existence, both individual and social. It leads also to virtualization of death and culminates, as I have shown elsewhere,[24] in the scenario of euthanasia for humankind.

II.4. Anthropological Scenarios and Projects for Modernity

Progressing chronologically, the historical description of anthropological formations naturally tends to turn into a forecasting discourse, a discourse of projects and scenarios. In the prism of philosophy this is one of the most dubious discourses, but in the situation of the present-day world, which is changing incomprehensibly, rapidly, and dangerously, it acquires significance and evokes interest. Taking this into account, let us attempt a cursory comparison of the forecasting aspects of the three projects that were touched on above: Deleuze's project, or the "form Superman"; Foucault's project, or the "esthetics of existence"; and synergic anthropology. All three projects take into account present-day reality and its leading trends, and therefore they agree as far as the most general matters are concerned. They are nonclassical projects that have fully separated themselves from all the European tradition of essentialist anthropology. They pinpoint radical changes occurring in the being "Man" and attempt to perceive the features of the one "who comes after the Subject."[25] Naturally, the features that they catch are different.

Deleuze's project, which we find in a short appendix at the end of his book on Foucault, is presented as only the briefest sketch. Its basic "form Superman" is not elaborated into a concept, although such elaboration he

24. See Sergey S. Horujy, "Euthanasia," in *Ocherki sinergijnoj antropologii* (Essays in Synergic Anthropology) (Moscow: Institut filosofii, teologii i istorii Sv. Fomy, 2005), pp. 110-24 [in Russian].

25. This is the title of a collection of articles and has become a popular formula: *Who Comes After the Subject?* ed. Edoardo Cadava, Peter Connor, Jean-Luc Nancy (New York and London: Routledge, 1991).

deemed necessary for any object of philosophizing. Nor does he relate it to his previous studies, although many of them seem to bear directly on the anthropological project: such as, for example, the elaboration of the concept of *visageité* (the constitution of the human face, "visageness") in *A Thousand Plateaus*, or of the concept of the brain (this elaboration concludes meaningly *What Is Philosophy?*). "The superman is defined by a new mode of feeling,"[26] he stressed already in his early book *Nietzsche and Philosophy*, but in the "form Superman" this aspect is also not developed. In short, this is rather the embryo of a project; nevertheless it is not only full of content but displays most characteristic features of Deleuze's thought. Of the three projects this one shows the greatest confidence in the course of things in the modern world, the greatest confidence in the prospects of technogenic civilization. Deleuze is full of enthusiasm with regard to the newest technologies, to the prospects of the manipulation of human genetic code and of transformations of Man into Cyborg: here he sees new and rich possibilities of interactions — not of man, no! — but of "forces in man" with new and diverse "external forces," with silicon chips, the bends of genetic chains, etc. The most important premise behind this enthusiasm is an immediately decomposing gaze that passes by man as such, does not see him, and immediately sees in his place "sensorimotor systems," "life, work, and language," "a membrane between the Internal and the External," and similar things, with a particular predilection for physico-biological and mathematical (topological) objects. Deleuze says: "One of our divergences with Foucault is that for him the social field is intersected by strategies whereas for us it entirely slips away."[27] This divergence originates mainly in the fact that for Foucault, especially in the late period, social reality is formed by the strategies of man, who develops certain practices of the self, a certain culture of the self; in other words, this reality has certain *irreducible authentically human dimensions*. But in Deleuze's discourse this would be impossible and unimaginable: in his discourse this reality represents "singular processes that take place in sets." In Foucault's work, the "culture of the self" and the "care of the self," the whole project of the practices of the self, signified, if not a breakthrough to man, then at least an impulse to such a breakthrough, a nostalgia for man, and when Deleuze describes this project, he does his

---

26. Gilles Deleuze, *Nicshe i filosofiya* (Nietzsche and Philosophy) (Moscow: Ad Marginem, 2003), p. 324 [Russian translation].

27. Gilles Deleuze, *Peregovory* (Negotiations) (St. Petersburg: Nauka, 2004) [Russian translation], p. 199.

utmost, without distorting it, to cleanse it of anthropological nostalgia. As we have already said, he (justly) underscores in various ways that Foucault's project is an investigation of modes of subjectivation that does not presuppose any construction of the subject; but, in addition, he conveys this project in terms of his own "topology of the fold," in which terms it becomes wholly Deleuzian, as a nomenclature of modes of subjectivation arising as "variations of changeable folds." One can clearly see in *The Hermeneutics of the Subject* (which, it is true, Deleuze did not know) that such a discourse is profoundly alien to Foucault's project. Returning now to Deleuze's own project, one can affirm that the problematic of the culture of the self and of the practices of the self is completely inappropriate to this project and is absent from it. Deleuze accepts without any doubts or hesitations the newest genetic and computer-based semi-utopias leading to the radically dehumanized Post-human (which he identifies with the Superman).

It would be fair enough to say that Deleuze's project is more topological and technological in nature than anthropological. In contrast, Foucault's project is undoubtedly an anthropological one and is not sidetracked in the direction of scientific and technological novelties; but, remaining in the anthropological sphere, it just as attentively takes into account the newest phenomena in this sphere and goes out to meet them just as decisively. Not less significant is another difference: if Deleuze observes and discusses the processes that occur, the anthropological phenomena that arise, Foucault chooses a more engaged position: he puts forward and defends his own model, his own anthropological scenario. A specific characteristic of this model is that it closely links the experience of antiquity with that of modernity, that it is at the same time grounded in the Greco-Roman culture of the self and connected with present-day extreme trends. Foucault finds the idea of the esthetics of existence in antiquity (although it would probably be safer to say that he attributes this idea to antiquity) and develops it using abundant materials from antiquity; then he argues that it is most topical for the present day and elaborates its application to the current anthropological and social situation. And the "neotribalist model" that results from this application of the ancient idea turns out to be just as extreme and decisive in accepting the most radical anthropological and socio-cultural trends as any movement or strategy with a post-human orientation. Moreover, in the case of Foucault it is a question not of accepting extreme trends but of forming, shaping, and guiding them. The role chosen by Foucault in his last years is more active and radical than that of Deleuze. In his last interviews — especially in those given to representatives of "subcultures" — his speech consciously avoids any commanding tones or intonations of a tribune, seems to avoid stressing or pressing; but nevertheless it is unmistakably the *speech of a teacher of life*. It was not for nothing that he got so used to Epictetus and Seneca.

As for synergic anthropology, we have already seen that its complex relations to Foucault's project are highly dualistic, including elements of closeness as well as of divergence. This dualism can be found in the forecasting aspects as well. An obvious element of typological closeness is that the linking of the experience of antiquity and modernity is also an essential feature of synergic anthropology: here, in the analysis of the newest metamorphoses of man, one retains the perspective of the Ontological topic and the orientation toward the experience of the ancient spiritual practices. In the characterization of the present-day anthropological situation there are no big divergences: the interpretation of this situation as the beginning of the period of Virtual Man's dominance contradicts neither Foucault's project nor that of Deleuze. The continuing expansion of virtual practices, the widening of their spectrum, the increasing importance of their role in human existence — these are all obvious and indisputable signs of the time. It is basically to the sphere of virtual practices that the "psychedelic subculture" belongs, to which Foucault assigns an important place in his project. In the "form Superman," "the forces of silicon taking their revenge on carbon" and transforming Man into a Cyborg demand man's transition into the Virtual topic (though the finale of the revenge, the Cyborg as such, is already beyond the anthropological reality). Like the appearance of the "new external forces" discussed by Deleuze and like the emergence of the "subcultures" propagandized by Foucault, virtualization is one of the leading trends forming the anthropological situation and its development; and each of the three projects, depending on its basic concepts and the logic of its ideas, puts at the center one or another trend of this sort. The only thing that is relatively new in our project is the conclusion that virtualization, when it becomes more deeply implanted, will inevitably turn into the scenario of euthanasia, the virtualized death of humankind. However, this conclusion too borders on the obvious (and one can note that the scenario of trans-humanism, the radical variant of the "revenge of silicon" dear to Deleuze, is also a form of euthanasia).

But each of the projects also has an aspect of evaluation; they all involve a certain position with respect to the situation and voice their preferences and recommendations on choosing anthropological strategies. It is here that profound divergences begin. Of the three projects, Deleuze's

Spiritual Practice, Synergic Anthropology, and Foucault's Project

position is the simplest and most straightforward: his choice and his recommendation to man is the experience of *transgression*, which must be realized with maximum fullness and perceived with maximum intensity (and the "form Superman" is conceived by Deleuze as maximum progress in this direction, as *superhuman* transgression). For French thought this has been a most familiar and perfectly worked-out position. Bataille and Blanchot have been its main teachers and promoters, and Blanchot's essay "Limit-Experience" (1962) formulated with great clarity what type of experience is involved: "We suppose that, in his essence, man rests satisfied: as universal man he has nothing to do . . . he reposes in the becoming of his immobile totality. Limit-experience is the kind of experience that awaits the higher man who is capable of going beyond such satisfaction . . . it is the experience of dissatisfaction of one who is satisfied 'in all things'; it is a deficiency, a pure defect, where, however, a culmination of being occurs. . . . Limit-experience is experience of this void at the edge of all fulfillment; it is experience of that which has place outside of everything when everything excludes all 'outside,' of that which remains to be attained when everything has been attained."[28] In terms of synergic anthropology, this is evidently extreme experience, but the maximum breadth of forms of the limit-experience to which Deleuze summons us certainly excludes ontological extreme experience; it remains within the limits of his topological reality devoid of ontology. Deleuze's position — and he gives it the character of a direct and insistent recommendation, rejecting in this case the role of an "abstract thinker"[29] — is the assertion of the Ontic topic.

Foucault's position is more complex and original. He too experienced the influence of Blanchot and Bataille, wrote the essay "On Transgression" (1963), and even in 1980 still said the following: "The idea of a limit-experience which uproots the subject out of himself — this is what was im-

---

28. Maurice Blanchot, "Opyt-predel" (Limit-Experience) in *Tanatografiya Erosa* (Thanatography of Eros), ed. S. L. Fokin (St. Petersburg: Mifril, 1994), p. 69 [Russian translation].

29. Cf. for example: "How ridiculous he is, this thinker. . . . When Bosquet speaks of the eternal truth of a wound, he is speaking from the name of a personal, disgusting wound which he bears in his own body. When Fitzgerald and Lowry speak . . . they are speaking from the name of all the liters of alcohol they have drunk. . . . When Artaud speaks . . . this is speech from the depths of schizophrenia. Each of them risked something and went here to the end; this gave them an indisputable right to say what they said. But what falls to the lot of the abstract thinker who gives wise advice . . . ? Is it not time finally to become professionals in these fields? . . . We should be a little bit of an alcoholic, a little bit crazy, a little bit of a suicide, a little bit of a partisan terrorist." Gilles Deleuze, *Logika smysla* (The Logic of Sense) (Moscow: Akademiya, 1995), p. 190 [Russian translation].

portant for me in my reading of Nietzsche, Bataille, and Blanchot, and what led me ... to always conceive [my books] as direct experiences aimed at uprooting myself out of myself, at preventing myself from being the same."[30] It is also important to note that such experience served for him as a concrete example of experience that did not conform with the phenomenological paradigm[31] and thus facilitated the emancipation from phenomenology, which was one of the essential tasks that shaped Foucault's evolution as a philosopher. However, in his late project he abandons extreme experience altogether, not only in the form of Bataille and Blanchot's limit-experience but also in all of its more fundamental representations, those, for example, that one finds in Kierkegaard, Levinas, and Heidegger. He refuses to base the constitution and hermeneutics of subjectness on extreme experience, although in one form or another, explicitly or implicitly, they always had to rely on such experience! Instead, he becomes a defender of the hopeless task of asserting a "purely immanent" human constitution in the spiritual exercises of his "Hellenistic model" (the exercises that did their best not to go too far and not to dig deep into the "self") and in the practices of the modern "esthetics of existence" based on this model. Nowhere does he say that he is rejecting extreme experience, but it is clear nonetheless: extreme experience, which by its definition is the experience of going out of oneself, experience that "uproots the subject out of himself," certainly cannot be an experience of pure immanence that "ties one to oneself — not to anything or anyone else except oneself" (581R; 514F); it cannot be based on the Foucaultian principle of immanence: "the unconditionality and self-sufficiency of one's relation to oneself" (581R; 514F).

Nevertheless, the rejection of extreme experience in favor of experience of the purely immanent does not succeed eventually. Foucault described fairly clearly (especially in the unpublished notes from which Gros quotes) what the purely immanent type of human constitution should be

---

30. Foucault, "Entretien avec Michel Foucault," *DE II*, no. 281, p. 862.

31. It is worth pointing out that in synergic anthropology the relation between extreme experience and experience in the phenomenological (intentional) paradigm is an essentially different one. The sphere of extreme experience is extended here thanks to the experience of the Ontological topic, which is recognized neither by Foucault nor by Deleuze as an autonomous form of extreme experience. Nevertheless it *is* such a form, and, as we have seen, the experience of consciousness in the mode of watchfulness that is included in it possesses the nature of intentional experience, while being at the same time extreme experience. But the cardinal distinction between the limit-experience and the experience of transgression, on the one hand, and intentional experience, on the other hand, is indisputable.

Spiritual Practice, Synergic Anthropology, and Foucault's Project

like; however, one can show that in the practices he considers, it is not this type that is realized, but something else. Disputing Foucault, Hadot demonstrated that, in the examples from the Stoics examined by Foucault, the constitution is in fact not immanent at all. Moreover, it is possible to affirm even on a more general level that a purely immanent constitution is a pure illusion, analogous to Baron Münchhausen's purely immanent extraction of himself out of a bog. If the "immanent constitution" of the subcultures described by Foucault is constituted by the "method for obtaining pleasure," then pleasure as such emerges for these subcultures as a purely transcendent principle.[32] Popular language catches this fact in the philosophically absolutely correct formula: the "*cult* of pleasure." I, however, am interested in the role and place of extreme experience. Experience that is not purely immanent does not by any means have to be extreme experience, and it is not hard to verify that Stoics' spiritual exercises represent exactly the kind of experience that is neither purely immanent nor extreme. Generally speaking, the experience of pleasure, too, is not necessarily extreme experience. However, in the project of the esthetics of existence this experience is chosen as the constitutive experience, so that pleasure becomes dominant in the configuration of all energies of man, and this means that it becomes a *passion*. Although this kind of passion is not as vividly extreme as transgression, nevertheless by its nature it, too, is a form of extreme experience, one that belongs to the topic of the unconscious.

But even irrespective of this conclusion about the nature of experience

---

32. Just as dubious is the "pure immanence" presented in Foucault's general schema of subjectification in Hellenistic practices: "Immanence establishes itself from self to self. All the exercises tend to establish from self to self a stable and complete relation, which one can conceive, for example, in a juridico-political form" (p. 581R; p. 514F). Let us ask: If the sought-for "relation from self to self" must be subordinate to a juridico-political form, is this form really immanent, found by the subject in himself? No, of course not: in the described subjectification it emerges as a transcendent factor, and the subject is constituted here not at all immanently but as a "juridico-political animal." However, the quoted lines are not Foucault's precise words but are Gros's paraphrase, and it is possible to suppose that the words about "juridico-political form" were carelessly added by him. But here are Foucault's precise words: "The self to which one has the relation is nothing other than the relation itself.... It is, as a result, immanence, or rather the ontological conformity of the self to the relation" (ibid.). As is easy to see, my question remains valid here too. That "pure relation" which is asserted here as constituting the *telos* of the practice of the self, is not arbitrary but a certain *necessary* relation. But where does the form of necessary relation come from? The source of this form is *transcendent* to the process of subjectification, and it is this source that is the genuine *telos* constituting subjectification. As for the relation as such, it turns out that it is not yet the real constituting factor.

described by Foucault, there is no doubt that the project of the esthetics of existence and of the "neotribalist" model is consonant with the most extreme anthropological trends. Foucault is very positively disposed towards the overall tendency of these trends, but in contrast to Deleuze he does not merely accept this tendency but attempts to "ride the tiger," to exercise a teaching influence on the anthropological and socio-cultural trends.

The position of synergic anthropology is of still a different kind. In the first place, its character is defined by the fact that it has at its disposal a complete description (albeit one that in this book has been sketched in cursory outline) of the types of extreme anthropological experience, and hence a complete repertory of paradigms of the nonpartializing human constitution. In surveying this anthropological panorama, synergic anthropology cannot fail to see in it a certain singular and singled-out point: the *topos of the Personality*. The source of its creation lies in a simple discovery: the sole mode of *actual* distinction from nonbeing is the Personality. Let me not dwell now on this thesis, although the above characterization of the Personality by means of the paradigm of *perichoresis* and the concept of "personal being-communion" does not yet by any means fully ground it. I will only note that the impersonal dynamics of the cosmos certainly does not constitute an ontological horizon that would be the Other to Nonbeing. Modern science, quantum field theory and cosmology, vividly shows how this dynamics describes mutual transformations of the essent and the nonessent, which together co-posit a single ontological mode and horizon. Further, the singled-out ontological role of the Personality implies the singled-out anthropological role of those strategies of Man in which in one way or another he establishes and actualizes his relation to the Personality. And finally among strategies of this kind a special and singled-out place belongs to the maximalistic strategy — that is, to a strategy which, without further ado and being motivated by a certain immediate impulse, sets for itself the (trans-)goal of actual transcending into the Personality, i.e., "deification." There follows immediately the question: *But is this possible?* Is not such a strategy nothing more than illusion, absurdity, madness? Before answering this question, it is worth indicating the specific characteristics of this strategy's *telos*: irrespective of whether it can be attained, this *telos* is an authentic *fulfillment* of man in being, in the ontological mode of perfect being-communion, and as such it is truly Man's maximal and highest idea of himself. If the defining characteristic of the Personality is actual ontological distinction from nonbeing, this *telos* presupposes also a certain radical transformation of Man's relations with the principles of nonbeing, nothing, and death. Profound connection of this *telos* with these relations is reflected in the moving impulse of the strategy in question: as we have shown,[33] its generating source is the "primary impulse of the non-acceptance of death," an impulse that enters into the fundamental structure of man's finiteness.

This maximalistic strategy is clearly nothing other than a spiritual practice oriented toward a personalist *telos*. Our discussion revealed its ontological roots and showed its special and singled-out role in the ensemble of all anthropological strategies. This affirmation of the singled-out role of a certain man's strategy is something that distinguishes synergic anthropology from other contemporary anthropological projects, including those of Deleuze and Foucault. *The position of synergic anthropology in problems of the evaluation of the anthropological situation and of the choice of strategies of man is formed with consideration of and in the light of the singled-out role of the topos of the Personality in the ontological aspect and of the paradigm of spiritual practice in the anthropological aspect.*

This means, first of all, that, in analyzing the anthropological situation and its trends, we pose an additional question, namely: How are they connected with the topos of the Personality and with the Ontological topic of the Anthropological Border? A general answer has already been given: this situation corresponds to the beginning of the dominance of Virtual Man while the leading trends strengthen this dominance; but this answer is far from sufficient here. As mentioned above, the spiritual tradition generates in the society a certain "adhering layer" around itself — a layer for whose members the world of spiritual practice serves as a reference system for their views, principles, and behavior. Therefore the fundamental structure of the Ontological topic includes not only the spiritual practice and the spiritual tradition as such but also a wide circle of strategies and practices "adhering" to the spiritual practice, i.e., oriented toward the latter and assimilating its most diverse elements to the most diverse degrees. (This whole circle of such strategies and practices has, until now, virtually not been described and not studied.) And the dominance of some other topic of the Border does in no way signify the complete disappearance of the spiritual tradition or of its adhering layer. Likewise, the leading trends do not by any means exhaust all the tendencies and possibilities contained in the situation, and in the highly dynamic anthropological reality they can change rapidly and be replaced by other trends. Although the major

---

33. See Sergey S. Horujy, *O starom i novom* (Of Things Old and New) (St. Petersburg: Aleteja, 2000), pp. 424-34.

features and tendencies of Man's situation are quite well determined and even, to a certain degree, predictable, the development of this situation is by no means *pre*-determined. The factors shaping the situation include man's will and his freedom, and, to one degree or another, the situation remains always fluid in character and accessible to influence. And as a result, the fate of the fundamental structures of the Ontological topic is also by no means predetermined.

In the light of the aforesaid, the principles forming the positions of synergic anthropology become clear. The latter's analysis of the anthropological situation must be more detailed, more intent, and directed not just at the new and dominant features but also at those still-preserved elements that are connected in one way or another with the Ontological topic. The data of the analysis, derived assessments, and recommendations must refer not only to the leading trends but also to these elements, to the perspectives of their existence. One should investigate the possibilities of the enhancement, extension, expansion of these elements as well as the ways and means of the creation of new elements of this sort (just us Foucault summons us to create and invent new modes of pleasure). The practices of Man must be "contemporaneous" to him and his situation, but who has ever told us that in the most contemporaneous and novel practices it is impossible to discover, liberate, and develop certain *resources of adhesion* that would make it possible not to lose our connection with the Personality? What helps the preservation of this connection in the present day is man's ability to creatively cultivate his personological dimensions, those dimensions in which he emerges as a being oriented toward the Personality. There are no guarantees that this connection will be preserved; on the contrary, in the present day it is in great danger. It is absolutely clear that cultivation of the personological dimensions and the preoccupation with the fate of the fundamental structures of the Ontological topic, with the fate of man as a being capable of becoming the Personality — it is absolutely clear that all this excludes unconditional trust in the present-day leading trends — trends that are leading to the euthanasia of Man, to cyborgs and mutants (euphemistically called "the revenge of silicon and genetic chains"), and to the breakup of mankind into minorities, seeking pleasure or choosing other kinds of un-ontological constitution. Here the divergence between synergic anthropology and the projects of Deleuze and Foucault is radical.

Foucault's project also includes a special and singled-out type of practices of man. He places at the center the "Hellenistic model" of practices of the self, finding these practices to be the apex of the Western culture of the self, regarding them as valuable for modern man as well, and even asserting that the revival of the ethics contained in them is a "task that is urgent, essential, and politically indispensable" (278R; 241F). Above I have spoken sufficiently about the roots and motives of this preference of his, of this "passion" of his (to use his own word) for the Hellenistic culture of the self. Now, in conclusion, it is time to ask another question: Can we follow Foucault here? Can we join ourselves to his preference and choice?

In order to answer this question, it is important to recall, first of all, that Foucault himself repeatedly points out the clearly unfinished character of the philosophical positions of the Hellenistic practices of the self, their indeterminacy and conceptual incompleteness in the key questions about man. That is the case even for the question of the meaning and destination of human existence, the question "What are the gods preparing man for?" As Foucault remarks, "[I]t would be very hard to find in Seneca an exact theory about this. . . . For Seneca this was not the important problem" (483R; 427F). Nor is another fundamental question, the ethical one, posed: "What does the division of people into good and evil represent?" Is this division innate and predetermined? Does retribution exist? And so on. "Neither Seneca nor Epictetus . . . seriously seeks an answer. . . . Both questions were strangely avoided by the Stoics" (483-84R; 427-28F). Further, he finds an analogous vagueness with regard to one such important issue of Hellenistic subjectivation as the definition of that "relation of the self to itself" which the practices of the self strive to achieve: "In Hellenistic and Roman thought it was never made clear whether the 'self' (*le* soi) is something to which one returns because it was present in advance, or whether the self is a goal which one must set for oneself and eventually . . . attain. Is the self the point to which one returns through a long ascesis and philosophical practice? Or is the self an object that one keeps constantly before one's eyes . . . ? Here . . . we have the case of the fundamental undecidedness" (240R; 205F) — and moreover this undecidedness touches on the key point, the constitutive *telos* of the practice of the self. It would not be difficult to extend the list of such philosophical "undecidednesses"; however, it is more important to note that the "Hellenistic model" turns out to be unfinished and incomplete in other aspects as well.

Foucault (whose views contain a perceptible residue of the French antireligious tradition) affirms not without satisfaction the religious indifference of the environment in which the Hellenistic model is cultivated. In the practices of the self included in this model the entire religious problematic is set aside and the religious experience is meager and shallow. The theme of salvation is present in these practices, but salvation is conceived not in a religious but in a philosophical discourse: it "functions like a philosophical concept in its own philosophical field.... In this concept of salvation which one finds in the Hellenistic and Roman texts, one does not find any reference to anything like death, immortality, or another world" (208R; 177F). Gros specially notes that "Foucault underscores the indifference of the Roman Stoics to questions of the afterlife."[34] The spiritual exercises themselves in these practices of the self "were not targeted at recognizing in oneself the divine principle" (456R; 402F), and even less so at achieving openness or the unlocking of oneself towards this principle. In sum, the religious dimension of the "Hellenistic model" must be recognized as mostly ignored and totally undeveloped.

Finally, the depth and the scale of self-transformation in the practices of the "Hellenistic model" depend critically on a key characteristic of this model, which is important for Foucault and strongly underscored by him: namely, they depend on the preservation in all these practices of an untouchable core of the "self" — a core that is subject neither to change nor to analysis or "decoding." This preservation entails a whole series of consequences. With the "untouchable core" is connected a particular sphere of human experience, which is thereby excluded from the work of comprehending self-perception and self-transformation. The experience of self is articulated not fully and not in all its depth; certain limits are placed on its processing and analysis. This is one of the reasons for the above-noted indeterminacy of the *telos* of the practices of the self, the vagueness of its nature and status. It is clear that, given these exclusions and limitations, the experience of these practices can certainly not be an experience of spiritual practice, an experience of human ascent and unlocking toward the ontologically Other. Indeed, such transcending experience involves in the activity of self-transformation the entire anthropological reality, all the levels of the organization of the human being; for this activity there can be no closed-off regions or untouchable enclaves in man. Because of these limitations this type of experience is certainly not an experience that gives a complete expression of some specific "nature"; accordingly it cannot possess an *organon* and its methodology cannot be philosophically funded comprehensively. All these properties are of the same kind; they express a very limited level

---

34. Frédéric Gros, "O kurse 1982 goda" (On the Lecture Course of 1982), in Michel Foucault, *Germenevtika subekta* (Hermeneutics of the Subject) (St. Petersburg: Nauka, 2007), p. 581R; p. 514F.

of the inner organization of experience, of its reflexive apprehension and epistemological grounding.[35]

In sum, in all its main aspects, philosophical, religious, and anthropological, the Hellenistic model of the practices of the self demonstrates the same character, unfinished and conceptually incomplete; the thought here fails to go to the very end, to the "last questions," and achieve comprehensive vision of its object — man, the "self." The model is based on an insufficiently mature and insufficiently profound experience of the "self." Man leaves unanswered here many essential questions about himself; he refuses to analyze himself in the whole scope and fullness of his inner world. His self-comprehension and self-transformation do not extend to certain spheres of the "self"; his relations with himself do not attain the ultimate depths. All this leads to a very definite conclusion: this model, which is imperfect and

---

35. There arises here an additional question: According to synergic anthropology, to which category of man's practices belong those practices of the self which satisfy Foucault's condition of preserving "an untouchable core of the self"? As said above, they cannot be spiritual practices, but a more definite answer depends on the content of the core. In the Hellenistic practices discussed by Foucault this core of the "self" does not include extreme anthropological manifestations, which means that their experience is not an experience of the Anthropological Border, and hence the type of subjectness constituted in them corresponds to a certain partializing constitution of man (the last conclusion coincides essentially with Kierkegaard's characterization of the Stoic practices). But in the version that Foucault suggests for modernity the main content of the core, the "core of the core," so to speak, is the method used to obtain pleasure, and as for the *telos* of the practices, by Foucault, "the goal is the experiencing of pleasure and its possibilities." This is already a different kind of experience, for such a goal definitely corresponds to passion, to man's passional organization, which is connected with the unconscious and shaped by anthropological manifestations from the Ontic topic. This means that the esthetics of existence in Foucault's modern version is anthropologically not identical with the ancient esthetics of existence since the latter, generally speaking, did not cultivate the passions (on the contrary, more often than not, it waged a combat against them) and did not establish subcultures based on them. Deleuze expressed similar judgments. He too finds in Foucault's last project a "summons to passion," and he notes the distinction between this project and the ancient culture of the self: "We do not return to the Greeks when we study the modes which are emerging in the present day" (*Negotiations*, p. 125). However, the similarity of our judgment with Deleuze goes no further. As we noted above, in spiritual practices the abiding in a passional state is judged to be an abiding in captivity, an enslavement of man — an infrequent case when the language of power is used in these practices. It follows that from the viewpoint of spiritual practices the "creation of the self" in the form of the invention of new and more exciting pleasures proclaimed by Foucault is a kind of slavish existence, and the death of Foucault himself from HIV (which at that time spread predominantly among homosexuals and drug users) is the death of an enslaved individual who has been punished by his master, passion.

unfinished in so many respects, could basically have been only a "temporary" and intermediate formation of the practices of the self. And here lies, to a significant degree, the explanation of its historical fate, which Foucault complains about — its later eclipse by the neighboring models, the Platonic and Christian ones. Further progress, a next step in man's relations with himself, was necessary and inevitable. The birth of some new formation of the practices of the self, where man's apprehension and processing of himself, of his consciousness, of his subjectness, would not stop at any limit but would strive to go to the end, to an exhaustive fullness taking in the whole content of the "self" — this birth was a logical and spiritual imperative.

Spiritual practice was this new formation — spiritual practice as we have described it, and not the phenomenon that Foucault represents under the names "the Christian model" or "Christian ascesis." I will return soon to Foucault's treatment of Christian ascesis, but now let me note that, in spiritual practice, man really did think himself and define himself to the end, to his own limits. The maximalistic task of actual ontological transformation has as its necessary condition a global and total self-examination (hermeneutics, exegesis, decoding, etc.) as well as an exhaustive and all-embracing self-transformation. In turn, practice of this sort requires a different type of subjectness: it can be accomplished only by a new "first-person subject." This subject is no longer an "ensemble of impersonal forces"; it is connected with and constituted from the horizon of personal being-communion. The transition to a new type of subjectness is reflected, in particular, in that *inversion of roles* which Foucault notices in his analysis of *parrhēsia*: if the old "self" was constituted by the Teacher's "true speeches," then the ascetic in the spiritual practice is constituted by the Source-beyond-there, the unlocking to which is realized through his own "true speech." The "first-person subject" organizes his own personal and unique relation with God, and for this he must articulate his experience himself and with maximum completeness. This is achieved, as we have seen, in "personal communion apropos of experience," and such communion is established by the spiritual tradition and its *organon*.

The spiritual practice in question is the hesychast practice — precisely the practice whose representative in the early stages was John Cassian. In its finished and mature form this practice becomes the foundation of the Eastern Christian discourse, whereas in the West the work of John Cassian was not continued and hesychasm did not take root there. On the contrary, as said above, the key hesychast conception of synergy was condemned in the West, and hesychast practice as well as hesychast energic theology have been rejected unreservedly until recent times. Foucault constructs a "history of the Western subject," of the latter's practices of the self, but the "Christian model" he includes in this history is Christian ascesis according to Cassian. His choice is not only understandable but even, in a certain sense, optimal: Cassian followed the ascetic path in the West and wrote in Latin, but at the same time he belonged to the hesychast line in the womb of which was conceived a comprehensive practical anthropology and "hermeneutics of the subject." Here was rich material, and the thinker used it with virtuosity (let us recall once more "The Combat of Chastity"). But in Foucault's construction Cassian's ascesis is torn away from its actual context, and there is confusion between the spirituality of the Christian West and that of the Christian East. In addition, although Cassian alone is analyzed, that which is constructed on the basis of the analysis is claimed to be the "Christian model of practices of the self," a reconstruction of Christian ascesis as such. And one must say immediately that this claim is unfounded, not just because Cassian was only one author and not even one of the major pillars of the tradition (as were, for example, Evagrius, the "Macarian corpus," Isaac the Syrian, and John Climacus) but also because he was writing at an early stage of the tradition, when many important elements of its *organon* had not yet been formed.

This loss of the true context led also to direct distortions and errors. Cassian and the practices described by him are part of the spiritual tradition, of a specific phenomenon constituted by the task of the faithful safekeeping and strictly identical reproduction of the experience of man's ontological ascent to Personality. This constituting task determines all the particular tasks and aspects of the life of the tradition, and they can be properly understood only in the light of it. *The only correct principle of the hermeneutics of the phenomena of spiritual practice is the principle of interpretation on the basis of the telos of the practice and its organon.* This interpretation should localize the phenomena in question with respect to the ladder of the spiritual-anthropological process of the practice. Foucault's treatment of Christian ascesis has nothing in common with this principle, however. All the fundamental concepts that form the necessary framework for understanding hesychast ascesis (and Cassian's ascesis in particular) are absent from this treatment including even the basic stepwise paradigm (the idea that ascetic practice has the structure of a ladder whose steps ascend from conversion and repentance to synergy and deification). To this is added the fact that he never mentions either Christ or prayer, even though it is impossible to exaggerate the role of both in ascesis. The absence of an integral image of the phenomenon, along with the extreme narrowness of the phenomenal base, leads to the fact that in the orbit of Foucault's analysis only a few scattered fragments of the phenomenon are present; and moreover, the absence of this integral image of the living whole to which these fragments belong leads to incorrect interpretations of them.

Of course, in Foucault's conception every practice of the self is goal-directed, and its goal determines many features of the process of the practice: this is an indisputable statement, analogous to the thesis of the constitutive nature of the *telos* of the spiritual practice. But in the case of the "Christian model" this goal, as Foucault constantly asserts, consists in the "total renunciation of the self"; and accordingly the path to this goal is nothing other than an increasing series of renunciations of newer and newer contents or parts of the "self," a series that steadily moves toward complete self-negation and self-devastation. It is strange to read all this, since it is so incorrect for a multitude of indisputable reasons. I will limit myself to only two reasons. The chief and already sufficient reason is that Foucault's assertion diverges — and even glaringly diverges — from the true state of things. Usually he makes this assertion not basing on any evidence, and in a few cases he demonstrates the presence of "self-renunciation" at the beginning of the practice, in repentance; but he never demonstrates the principal thing — the series of renunciations continuing over the entire course of the practice and attaining its apogee at its highest steps. This cannot be demonstrated in any case, because it does not exist. The highest steps of the hesychast ladder are virtually absent in Cassian's discourse: in his days they were still poorly elaborated. As for Foucault, his texts leave the impression that he simply does not know anything about these steps, not even their names. But it is precisely here that the most unique things take place which single out spiritual practice as the *ontological* practice among all other practices. As described above, this is where the spontaneous creation-generation of new anthropological energic forms begins, the forms whose real existence is confirmed by an abundant corpus of experiential evidence and reflected upon in a rich ascetic and theological discourse. There take place the establishment of the "mind-heart" and the formation of the ontological mover, the coupling *prosochē-proseuchē*,[36] the generation of "noetic feelings" and much else; and in the light of all this subtle personalistic building compared to which the Stoics' exercises are (to quote Chekhov's hero) like the work of a rough carpenter compared to the work of a skilled wood-craftsman — in the light of this, what can one say about Foucault's assertion? It is best to say

---

36. "Attention-prayer." — Trans.

nothing. But I still planned to indicate the second reason why his assertion is glaringly incorrect.

Here is what is incomprehensible and strange: the paradigm of successive self-renunciation (of ever-deepening dismantling, dissolution, leveling of personal structures, self-devastation, and so on and so forth) that Foucault attributes to Christianity, has always been known to religious and philosophical thought and is firmly linked with a specific, not at all Christian type of mysticism and mystical practices. As Foucault could not have failed to know, this paradigm exactly corresponds to the *haplosis,* or "simplication," of the Neoplatonists; it also plays an important part in the impersonal practices of the Far East (where it is perhaps most vividly expressed in Taoism). Christianity is at the opposite pole from this! The Christian in his spiritual life — in particular, the hesychast in the practice of ascesis — strives to attain not the Great Void, but Christ; and his striving is realized in the element of personal communion, which tries to become intensified up to the interchange of being, the *perichoresis.* The process of the practice here takes place in the element of gain, not of loss; in the element of self-building, not of self-dissolution. Even the necessary presence of the self-renunciation at the initial stage signifies not a loss of identity and of the continuity of personal consciousness, but their potentiation, to use Kierkegaard's term: "Peter remains Peter, Paul remains Paul, and Philip remains Philip; each of them, having become filled with the Spirit, abides in his proper nature and essence."[37] That is the voice of the classical hesychasm of the fourth century, and in the twentieth century, tradition expresses the same experience: "The Person-Hypostasis is revealed within the penitent. . . . The hypostatic principle is actualized within us."[38] In the same city of Paris where Foucault read his course, there appeared and became widely known a book by Vladimir Lossky which explained that in Christian ascesis "human personalities are by no means brought into any process that could abolish their freedom and destroy their personhoods."[39] These quotations are sufficient to enable us to evaluate the merits of Foucault's position.

---

37. Macarius of Egypt, "Nastavleniya o khristianskoy zhizni" (Instructions Relating to the Christian Life), in *Philokalia*, vol. 1 (Sergiev Posad: Svyato-Troitskaya Sergieva Lavra, 1992), p. 274 [Russian translation].

38. Archim. Sophrony Sakharov, *Videt' Boga kak On est'* (We Shall See Him as He Is) (Essex, UK: Stavropegic Monastery of St. John the Baptist, 1985), pp. 164, 177 [in Russian].

39. Vladimir N. Lossky, *Ocherk misticheskogo bogosloviya Vostochnoy Cerkvi* (The Mystical Theology of the Eastern Church), Bogoslovskie Trudy, vol. 8 (Moscow: Izdatelstvo Moskovskoi Patriarkhii, 1971), p. 97 [in Russian].

Given the overall unsoundness of this position, it is not that important how it presents some or other individual themes; I will touch only upon the most principal ones. Undoubtedly, for Foucault one of them is repentance. His analysis of it is based virtually on a single source: Tertullian's treatise *On Repentance;* but besides this paucity of the phenomenal base, another thing, no less important, should be noted. Foucault could not have chosen a worse source if he wants to consider the phenomenon as the practice of the self and the aim of his analysis is to reconstruct the structure and processes in the penitent consciousness in order to outline the anthropology and subjectology of repentance. And his source is bad not just because of its early date, thanks to which it could have no connection with Christian ascesis (which had not yet been born) or with the mature and developed culture of Christian repentance. More important than this is the fact that Tertullian, a stylist and rhetorician of genius and a brilliant apologist, possessed only a crude and superficial apprehension of psychical and spiritual life and was by no means a discoverer of depths of consciousness. What the penitent consciousness is, is something he did not plan to explain, and had he wished to do so, he would not have been able to. This is how his treatise on repentance is appraised by an acute and profound spiritual writer, the church historian Fr. Sergii Mansurov: "The writer of genius turned out to be a shallow Christian. . . . What strikes us first of all is the disharmony between the severe demands of 'spirituality' . . . and what he says about the positive content of spiritual life. He speaks most of all and most eloquently only about the outer shell of this life . . . as if it were the essence of the matter. . . . How pale is that which he says in *On Repentance* about the inner regenerating effect of repentance. There is a compulsion to repentance here . . . but there is no description or disclosure of its essence."[40] Mansurov judges the treatise to be an "unsuccessful and shallow essay," in which the author showed his "ineptness in describing the inner Christian life." As is clear from the foregoing, Tertullian's treatise, which puts forward nothing more than "severe demands" and a "compulsion," fits Foucault's "Christian model" perfectly, and this is sufficient for him to place this "unsuccessful and shallow" text, in which "there is no description or disclosure" of the essence of repentance, at the foundation of his own conception of repentance.

It is surprising, however, that, in spite of its source, Foucault's concep-

---

40. Fr. Sergii Mansurov, *Ocherki iz istorii Cerkvi* (Sketches from the History of the Church), *Bogoslovskie trudy,* vol. 7 (Moscow: Izdatelstvo Moskovskoi Patriarkhii, 1971), pp. 109-10 [in Russian].

tion was not blind to the essence of repentance as an *ontological* event, as a fundamental choice of an ontologically alternative anthropological and meta-anthropological strategy. He conveys this ontological nature of the event vividly and expressively: "In this Christian metanoia — in this sudden, dramatic, happening within history and yet metahistorical collapse of the subject, you deal with a transition: a transition from one type of being to another, from death to life, from finite being to immortality, from darkness to light, from the kingdom of the devil to the kingdom of God, and so on" (237-38R; 203F; I am quoting this excellent text a second time). Nevertheless, it was inevitable that there would be numerous errors in his treatment of repentance. As an example, let us indicate one of his major errors. An insistent leitmotif of the ascetic discourse is the necessity of *unceasing repentance*. This necessity is connected with many key properties of the hesychast Ladder (the very existence of which is ignored by Foucault). It is sufficient to say that the energic nature of ascetic ascent implies the necessity of an unceasing effort, of "doing" (praxis), which at each of its steps preserves and reproduces concisely all the preceding steps. For this reason "we have need of repentance all twenty-four hours of the day and night,"[41] as the most authoritative teacher of ascesis instructs us. In contrast, Foucault identifies "the meaning introduced by Christians" into the concept of repentance with the meaning given to repentance by the late Pythagorean Hierocles,[42] who defined repentance by the formula "preparation for life without pangs of conscience." Varying this formula, Foucault says that Christian repentance "opens access to a life where there is no place for penitence" (242R; 207F). But that is grossly incorrect. The life about which he speaks does not exist for the Christian, and repentance opens access for the Christian not to that life but to the Ladder of ascent, where he will repent "all twenty-four hours of the day and night."

In conclusion it is worth revisiting once more the historical and comparative theme of the relations between the Hellenistic and Christian formations of practices of the self. Describing the "Christian model" in Section I, we noted that, on the main level, Foucault presents these relations in a simple and straightforward way: he interprets all the common elements of the two formations as being due to borrowings by Christianity. But then we

---

41. Isaac the Syrian, *Slova podvizhnicheskie* (Ascetic Orations) (Moscow: Pravilo Very, 1993), p. 201 [Russian translation].

42. This author of the fifth century CE should not be confused with the aforementioned Stoic of the first to second centuries.

have shown that the Christian practices of the self developed the cardinally new paradigm of spiritual practice, and its basic principles — the ontological ascent to Other-being as the Personality and the complete *organon* of the experience of ascent — were such that it was impossible to borrow them from anywhere; they could only be totally new discoveries. On the basis of these principles there arose a new anthropological process unknown to the man of antiquity and a new anthropological context was gradually forming up; and this process and context also could not borrow anything essential: their specific nature made it impossible that something important would be "brought in from the outside and used as it is." Inevitably and indisputably, the world of spiritual practice had many elements in common with the ancient spirituality and culture of the self, but *the telos of the Personality and the Christocentric context* which determined everything in this world categorically excluded a simple transfer of these elements; they required that each of them undergo a certain radical "remelting." This term refers to Humboldt's epistemological metaphor of the "melting pot," which was already applied by Georges Florovsky to the creation of basic Christian concepts. (Cf.: "The fund of ancient words turned out to be insufficient for the theological Credo. It was necessary to reforge ancient words, remelt ancient concepts."[43]) A classical, paradigmatic example of such remelting was already mentioned above: it is the birth of the Hypostasis from two unexpected, mutually alien components — Aristotle's "first essence" and the "mask," *prosōpon*, the term not belonging at all to the philosophical discourse. In the crucible of the previously accumulated experience of the Personality, both components changed unrecognizably, and together they became the Person-Hypostasis: the notion that not only had a new content but was the concept of a new kind that became in turn the foundation of a new kind of discourse, dogmatic theology. This authentically Christian discourse, which originated in the fourth century with the decisive participation of the experience of the Councils, gave Christian thought its own grounds, language, and order as well as its own criteria, on the basis of which it carried out a global reexamination of all the extensive borrowings (this is the only place where this word is appropriate!) made from ancient philosophy in the early epoch. The fate of Origen's heritage is the best example of how profound and strict this reexamination could be, but Foucault, when speaking about the Christian borrowings, avoids mentioning the very fact of it.

43. Georgij V. Florovsky, *Vostochnye Otsy IV veka* (Eastern Fathers of the 4th Century) (Paris: YMCA-Press, 1931), p. 75.

Spiritual Practice, Synergic Anthropology, and Foucault's Project

As for practices of the self, the hesychast practice in its formation exhibited elements of closeness to the Stoic practices of the self insofar as the latter developed an integral vision of man and a tendency to organize the experience of self-transformation into a coherent system of exercises. It also exhibited elements of closeness to the Neoplatonic practices, for the latter embodied an ontological vision of man (though it corresponded to a different ontology). But at the same time the constitutive elements of this practice, the *telos* of the Personality and the Christocentric context, emerged as two fundamental factors that fully guaranteed its creative independence as well as transformative and remelting power with respect to all the phenomena taken into its organism. In sum, which model of relations can be seen in this situation? We are concerned only with anthropology now, and with history conceived as the history of man's relations with himself; and leaving aside all other dimensions, we can speak of the *creative succession of anthropological projects*. We have seen that in many of its significant aspects the "Hellenistic model" is unfinished in character, whereas in the spiritual practice a specific type of man's practices and a specific approach to the constitution of man attained completeness and received its final form. This makes it possible to see in the spiritual practice a kind of "fulfillment-transcending" of the ancient practices of the self. A retrospective view shows that many anthropological discoveries of the spiritual practice such as the *organon* of the experience and the community creating the *organon* (spiritual tradition), the hierarchy of the forms of communion, and so on, were already present in the womb of the ancient practices of the self as embryonic possibilities and emerging tendencies. In the spiritual practice these possibilities were fully actualized and acquired a mature form that it was impossible to guess in advance, and needless to say such fulfillment-transcending was extremely far from being a borrowing. But the relations of the creative worlds are profound and multidimensional; there is a place in them for pure borrowing (of not particularly important details), for the diametrical opposition of principles, and for ideological hostility. But, unfortunately, such a depth, scale, and dramatic character of the relations are not conveyed in any way by Foucault's reconstruction of the "Christian model" with its simplistic speech about "borrowings."

However, in evaluating Foucault's "Christian model," it would be unjust not to mention once more its indisputable success: the "brilliant essay" (as Peter Brown calls it) analyzing the passion of fornication. As an investigator of fornication, Foucault demonstrates the highest competence and scrupulousness.

To sum it all up, I come to a reception of Foucault's project that might seem discouragingly simple or even primitive. As far as the idea of the practices of the self is concerned, it merits nothing but the most respectful praise; in our opinion, this idea, which is revolutionary with respect to classical anthropology, is extremely relevant, extraordinarily rich, and exceptionally full of promise. Moreover, it has a whole web of connections with synergic anthropology, and these connections are useful and valuable for the latter. However, two peripheral, nonphilosophical factors have left a powerful impression on its embodiment: Foucault's ardent and militant homosexuality and his attachment to the long line of French secularist anticlerical and antireligious ideology. These factors implicitly influenced his anthropological positions and many of his historico-cultural preferences and judgments, while explicitly and directly influencing the modern anthropological strategies proposed by him. This influence, both implicit and explicit, was, in our opinion, far from being philosophically fruitful.

III

So Where Shall We Sail?

There is no doubt that the comparative discussion of the two anthropological projects I have undertaken in this book, despite all the divergences, shows them to be similar and close in their major features (thus confirming the first impression described in the beginning). It could not be otherwise, since such a great similarity and closeness are already implied by the identical initial position of the two projects, namely by the decision to take *anthropological practices* as the basic concept of the anthropological discourse. This decision presupposed a position of principle that accepts that *Man is constituted in his specific practices;* and this position, in turn, predetermined a new direction for anthropological discourse that was different from the known types of anthropology. The emerging paradigm of human constitution entailed a rejection of the classical essentialist anthropology of Aristotle-Descartes-Kant; the grounding in the practices of man signified an abandonment of the conceptual systems of philosophical anthropology, of this entire direction; at the same time the conception of anthropological practices did not coincide with any versions of the approach based on man's acts and activities and cultivated in neo- and post-Marxist thought; and so on. The initial position determined also the basic class of practices capable of constituting man. This class was identified unambiguously: it consisted of practices in which man occupied himself with goal-directed self-transformation. At first I called them practices of autotransformation; then, deciding that Foucault's term, practices of the self, sounds more expressive, I also adopted it.

For non-Russian readers: the title is taken from the final line of Pushkin's *Autumn*, one of the best Russian poems.

The final goals of both projects are connected with modernity; the principal necessary task was seen in the understanding of present-day man and the determination of his strategies. However, the "practical" direction had to form its own foundations — to define its phenomenal base, to develop methods, and to construct concepts. Both projects stated that Modern man essentially neglected practices of the self, and therefore it is only the epochs of the past that could provide the phenomenal base for the study of constitutive practices of man. As a result, the emerging direction acquired a specific doubly oriented structure: it had to identify in history and investigate a certain formation of anthropological practices — with the further aim to develop, on the basis of these practices, a new nonclassical anthropology that would enable us to understand the anthropological situation of the present day. As we have seen, this bidirectionality (toward past epochs and toward modernity) in virtue of which the ancient practices become a source of new concepts and ideas capable of solving present-day anthropological problems — this bidirectionality is a general characteristic of both projects. In addition, both projects trace the historical sequence of anthropological formations, of types of subjectness determined by types of practices of the self. But as one goes further, there are cardinal differences. If Foucault's project chooses as its phenomenal base the practices of the self of late pagan antiquity, then synergic anthropology places at the center the ontological paradigm of constitution, the unlocking of man in being, and accordingly it places at the foundation of its phenomenal base the practices that effect such unlocking, i.e., spiritual practices. This produces a whole spectrum of further differences and divergences, of greatly varying depth and acuteness.

Toward what tasks does this picture direct us? What is next? Of course, both projects still have unexhausted possibilities; they also have their own problematic and will, in all probability, evolve further, developing this problematic. But this joint examination of these projects leads to some broader questions — questions relating to the problems and prospects of the whole direction that has been formed in these projects. What is the principal experience that has been achieved by this direction, by this type of anthropology, which is radically nonclassical, essenceless, and subjectless? And where and how can or should this type of anthropology go?

As far as the experience is concerned, I would briefly characterize it as the *experience of a new anthropological landscape*. The experience of Man, of anthropological reality, has become completely different in character. Previously, in the good old days, this was the experience of a particular existent to whom anthropological thought continued seeking out the most fitting name (mortal? subject? person? individual?) and could not stop deciding which properties are indispensable parts of his essence. But today our experience reveals anthropological reality to be a certain strange landscape in which a multitude of peculiar post-subject anthropological structures are scattered about: "modes of existence," "scenarios of subjectivation," "energic forms of the Anthropological Border," etc. Deleuze's philosophy is also one that forms up this landscape, populating its far left flank with almost totally dehumanized topological-physical force formations. The landscape is not yet complete: the contribution of synergic anthropology must include anthropological structures that correspond not only to the topics of the Anthropological Border but also to the "adhering" strategies and practices of man, and these structures are not reconstructed so far (although they can be quite important for the development of the situation). Thus in the place of man we see a highly variegated collection all the members of which indisputably represent something "related to man," but it is just as indisputable that this collection is something very distant from the fully dimensional and full-fledged man that we had in the previous existent; in the place of a single integral existent, a large set of partial formations has appeared.

But what matters is not only that the formations are partial and truncated into various strange shapes. In any of the new anthropological structures not only do we fail to detect many of the customary predicates of anthropological reality, but we find the remaining basic predicates in a drastically changed form. A process of deep changes has overwhelmed the whole sphere of man's relations with his body, and the character of the process attests to an acute crisis of these relations. Man experiences disorientation as it suddenly becomes clear to him that he does not know anymore what his body means for him, and he starts seeking anew the meaning of his corporeality, subjecting it to extreme tests: taking it apart and cultivating all kinds of extreme bodily practices (in which many radical movements in modern art actively participate). A new culture of perceptions emerges, a restructuring of the relations of the modalities of perception, a cardinal increase of the role of the aural impressions, and so on. Some of the new "modes of existence" are oriented toward the restructuring of consciousness, toward altered states of consciousness (using various technologies, with or without drugs); in others, predicates of the usual mode are subjected to virtualization. One should specially mention changes that take place or are soon expected on the level of man's biological foundations: new techniques of human reproduction, manipulation of genetic material, cloning, and so on. Here the scale of the changes is potentially unlimited, to the point where Man as a genus disappears.

Finally, one of the new modes requires a separate discussion. It requires it because it is at the same time a new mode and the oldest one, based on one of the fundamental anthropological relations: *the relation with death*. In the majority of the new anthropological structures this relation underwent a profound change. A pluralization occurred here too: from having the status of "Man's fate," a unique and universal fate lying outside the sphere of man's will and activity, death became one of the scenario subjects, an object that one is to decide and experiment on. Deleuze and Foucault both actively defend the right to suicide; for them it is a sine qua non, one of the human being's inalienable rights.[1] Both thinkers develop the theme of an *apologia for suicide*, connecting this theme with the central problem of subjectivation and associating with suicide a special mode of existence. But they present somewhat different views of this mode. According to Deleuze, "the line of the external" embodied in practices of power "moves, possibly, toward death, suicide"; however, subjectivation (in Deleuze's discourse it is "an operation forming a fold on the line of the external") reveals "the only way to counteract this line and even to saddle it." And when an individual arranges a mode of existence capable of saddling the movement toward death, "suicide can become an art which will require all of life."[2] Such an idea of suicide is basically just a variation on the theme of "running ahead *(Vorlaufen)* into death" in *Being and Time*, but the instrumentation of the idea is very different in the two thinkers: although in both cases one distinctly hears the motive of the heroization of death, Heidegger, still a young man and in love with Hannah Arendt, specially stresses that, in this "running ahead," an individual is maximally far from all attraction to empirical death. As distinct from both Heidegger and Deleuze, for Foucault the suicidal mode of existence is arranged not in the key of severe heroics but in the key of pleasure. His principal text on this theme is a lyrical essay in praise of suicide with the alluring title "Such a Simple Pleasure." This

---

1. Cf. for example: ". . . finally the right of everyone to kill himself whenever he wishes, and in decent conditions, must be recognized. . . . If I ever won a few billion in a lottery, I would establish an institute where people who wished to die would come to spend a weekend, a week, or a month in pleasure, taking drugs perhaps, and afterwards they would disappear, as if they were erased . . ." Foucault, "Un système fini face à une demande infinie," *DE II*, no. 325, p. 1201.

2. Gilles Deleuze, *Peregovory* (Negotiations) (St. Petersburg: Nauka, 2004) [Russian translation], p. 150.

So Where Shall We Sail?

title is not only alluring, it is also significant: in Foucault's discourse of the late period, "pleasure" is the supreme principle of his anthropological utopia, and the inclusion of suicide in the category of pleasures means that Foucault assigns to it a certain mode of subjectness and a certain subculture which is also included in the ensemble of subcultures of the "esthetics of existence." Here are some of the principles of this mode that openly contradict Plato's dogma of philosophy as a preparation for death: "I am a little irritated by the wisdom which promises to teach people how to die and by the philosophies which tell people what to think about this. I am left indifferent by the fact that it is forbidden to us 'to prepare this.' One must prepare it, arrange it, fabricate it piece by piece, calculate it; one must find the best ingredients, one must use one's imagination, one must choose, seek advice, and work it to create out of it a work without any spectators, which exists for me alone and lasts for the briefest second of life. . . . We have a chance to have at our disposal this absolutely unique moment . . . to make of it an immeasurable pleasure the patient preparation of which will illuminate all of life."[3] Perhaps it must also be added that here, in the sphere of relations with death, not only the texts of Foucault and Deleuze but also their personal fates represent significant factors in the present-day anthropological landscape. These fates — suicide, death from AIDS — are organic parts of this landscape, for these are not the fates of "subjects" but the realization of modes of subjectivation belonging to the two thinkers or — to use Foucault's formula (which, applied by everyone to him, has become a cliché) — belonging to their lives as works of art. Given the fact that the realization was not virtual but actual in the highest degree, we can say that these were works of modern, rather than postmodern, art.

The importance of this theme for both Deleuze and Foucault compels us to draw some conclusions. What the two thinkers present here is their common *(en grand)* existential and philosophical position, and this position has its history in Western thought. Suicide was preached by the Cyrenaic philosopher Hegesias, called the "Teacher of death." Although he had success and gathered around him a circle of followers, the "Hegesians," his preaching would be virtually forgotten today if his name did not keep appearing on pages read and reread by generations of French intellectuals. In Flaubert's *Temptation of Saint Anthony*, Death says: "The philosopher Hegesias in Syracuse preached me so eloquently that people would abandon lupanars and run away into the fields to hang themselves." Hegesias'

---

3. Michel Foucault, "Un plaisir si simple," *DE II*, no. 264, pp. 778-79.

teaching is not that much older than the Hellenistic practices studied by Foucault, and our quotation from *The Simplest of Pleasures* makes it already clear that the position of Foucault (and of Deleuze) can justly be called a *neo-Hegesianism*, which complements their *neotribalism* in the sphere of the relation to death. In the culture of modernism and postmodernism the appearance of this remake of Hegesianism is just as natural and organic as the appearance of Hegesias himself in the twilight of antiquity. There is also the direct source of the remake, and it is, of course, once again Nietzsche, who wrote: "If you annihilate yourself, you do a deed worthy of the greatest respect."[4] From the positions of synergic anthropology the cult of suicide is one of the expected and quite significant signs of the scenario and trend of the euthanasia of the genus anthropos.

As yet I do not clearly see what the figures in the new anthropological landscape are composing themselves into. The landscape does not appear to be meaningful, and our gaze does not find in the overall picture any unifying principle. As far as I can see, the only qualities that unify the different anthropological structures in the landscape are the negative qualities of essencelessness, post-subjectness, and so on. Evidently, this lack of meaning in the anthropological landscape is exactly the next problem that needs to be tackled. Contemporary philosophy must achieve a meaningful vision for which the anthropological reality in all its fullness would represent not just a landscape but would be organized into an integral *anthropological perspective*. "Perspective" should be understood here in the epistemological sense, as *cognitive perspective*, as a way of seeing that includes a paradigm of knowledge. Both Foucault and Deleuze find that the phenomenological paradigm is, if not totally inadequate, then at any rate insufficient for the new philosophical and anthropological reality,[5] and the logic of the development of the new anthropology naturally leads to the point of view that this anthropology must create its own, properly "anthropological" (anthropologically funded and oriented) way of seeing, its own "anthropological" (in the same sense) cognitive perspective. Including phenomena into itself, this perspective will group them and establish their relations in its own way, and the historico-cultural context will look different in it. Let me present a small example showing that elements of this sort of perspective are already

---

4. Friedrich Nietzsche, *Sumerki idolov* (Twilight of the Idols), vol. 2 (Moscow: Mysl', 1990), p. 612 [Russian translation].

5. Cf. for example: "The phenomenological subject was disqualified a second time, by psychoanalysis, just as it had been already disqualified by linguistic theory." Foucault, "Structuralisme et poststructuralisme," *DE II*, no. 330, p. 1254.

implicitly present in anthropological projects considered. I have observed that Foucault finds in Epicureanism the specific phenomenon of "vertical translation," an organism of the identical transmission of the constitutive experience of the practice of the self. I have also pointed out that this phenomenon represents an embryo of the "spiritual tradition" in the sense of synergic anthropology, i.e., a crucially important element of the spiritual practice and, in particular, of the hesychast practice in Christianity. Foucault's analysis of Epicureanism identifies also another, not less essential, element of convergence with Christianity: a new culture of personal communion was maturing in Epicureanism that would become soon one of the principal consequences of the "discovery of personality" in Christianity. According to Foucault, a special "openness of the soul and heart in communion" is cultivated in Epicureanism, and it is precisely this openness which represents a prototype of Christian confession: "Here we encounter for the first time the duty we will later encounter in Christianity: to the word of truth which teaches me truth and thus helps me to be saved, I must . . . respond with true speech in which I reveal before another and before others the truth of my own soul" (423R; 374F). Both of these important elements of convergence have a specifically anthropological character, whereas the standard historico-cultural assessments based on ideological criteria, on the comparison of doctrines, see in Epicureanism a movement most remote from Christianity and a worldview polarly opposite to Christianity. Thus, Foucault's analysis organizes an anthropological perspective in which phenomena that are very far apart in the standard ideological perspective turn out to be close to each other.

It is on this pathway that the development of such a new perspective can be conceived. The elements of an anthropological perspective must be systematically identified in both projects; they must be scrupulously collected and put together. The conceptual apparatus of the projects is determined by the working principle of this perspective, which can already be seen in our example: phenomena of anthropological reality are included in the perspective by means of their description on the basis of practices of the self and (in the case of synergic anthropology) on the basis of practices of the Border. In other words, the anthropological perspective achieves a comprehensive vision of the anthropological landscape, representing it as formed by certain practices of the self. To a certain degree such a hermeneutics of the anthropological landscape is analogous to the "hermeneutics of the subject" already performed in both projects; however, the problem that now arises is of the "next generation": if previously it was a question of the constitution of separate subject formations (limited moreover in Foucault's project by the frame of "a millennium of care of the self"), the problem now is to achieve a unified vision of the entire extremely pluralistic, fragmented, mobile, and contradictory panorama of the anthropological reality, together with its trends, at the beginning of the third millennium. There are reasons to suppose, however, that the methodological and heuristic resources of the approach of practices of the self — or, in the terms of synergic anthropology, the approach based on the extreme anthropological manifestations — are sufficient to solve this problem.

They can do even more than that. The practices of the self, or the practices of the Anthropological Border, are constitutive for man, and the anthropological perspective formed on the basis of them leads us naturally beyond the frame of the anthropological landscape, or of the anthropological reality as such. In principle, it involves the entire "human-dimensional" reality, i.e., reality formed and organized by man's presence and activity. In fact, the arrangement of the historical sequence of anthropological formations that I have described above can already be viewed as a kind of inclusion of history into anthropology: it is a transformation and reinterpretation of the historical discourse as a result of which history is presented as the *history of Man*. Then this representation of history can and must be extended to include all related histories of the social, cultural, and other dimensions of Man's existence. Accomplishing all these transformations of the historical discourse, the anthropological discourse (represented as the discourse of practices of the self/practices of the Border) plays the part of a *meta-discourse* in relation to the latter. There arises the natural hypothesis (which, of course, needs to be substantiated) that it is capable of playing an analogous part in relation to other humanistic/human-dimensional discourses as well. One can find many arguments in favor of this hypothesis; in particular, the tendencies to such an "expansion" of the anthropological discourse can easily be seen in all the representations of this discourse in terms of practices of the self/practices of the Border.

Here is a good example from Foucault's theory. Describing how conversion to the self and knowledge of the self are realized and combined in Stoic practices, the philosopher arrives at the following pithy formula: "to direct one's gaze at oneself, while at the same time embracing the whole universe with it" (286-87R; 249F). He also expands it: "The Stoics ... insist on ordering all knowledge according to *technē tou biou*; it means to direct one's gaze at oneself, at the same time considering this conversion and this directing of the gaze at oneself as the possibility to embrace with this gaze the whole universe and see its general order and internal organization" (286-87R; 249F). Here conversion to the self, an authentically anthropological principle or paradigm, is reinterpreted as a complete *program of world-knowledge integrated into the anthropological problem and realized in anthropological discourse* (insofar as, in this "seeing of the order and organization of the universe," man's gaze remains directed at himself!). Such a new and generalized form of the paradigm of conversion to the self in Stoicism does not remain an empty claim: according to Foucault, it is embodied in Seneca's *Naturales quaestiones*, a work of encyclopedic character, "a great book which surveys the whole world" (288R; 251F). Evidently, this form contains a specific epistemological position that demands an expansion of the anthropological discourse, an anthropologizing transformation of all the discourses that participate in the description of the order and organization of the universe. Ideas of this sort are not difficult to find in Foucault himself: for example, as Gros shows, Foucault in his last period outlines an "ethopoetic" conception of truth — "a kind of truth ... which would have been read in the pattern of accomplished acts and bodily gestures."[6] In our logic, this conception presupposes the idea of the anthropologization of truth, which implies, in turn, the anthropologization of the cognitive paradigm. Once again, the expansion of the anthropological discourse is at work.

As I have already pointed out, in this expansion the anthropological discourse begins to play the role of a meta-discourse transforming the adjacent discourses and their relations. It thus acquires a certain new status. The anthropological perspective, which integrates into itself — potentially — the entire ensemble of humanistic/human-dimensional discourses and which moreover is endowed with its own epistemological paradigm, this perspective becomes the unifying principle and general methodological basis of this ensemble. It becomes the epistemological formation that embraces all this ensemble, and such formation can be called an *epistēmē*.[7] In the methodological and heuristic aspect the "anthropological *epistēmē*" is a radically new mode and status of anthropology but, in the general plane of the progress of ideas, its emergence does not have an extreme revolutionary character. As Gadamer observed, it was already in the period between the

---

6. Frédéric Gros, "O kurse 1982 goda" (On the Lecture Course of 1982), in Michel Foucault, *Germenevtika sub'ekta* (Hermeneutics of the Subject) (St. Petersburg: Nauka, 2007), p. 576R; p. 509F.

7. Needless to say, we have in mind here the direct "etymological" sense of the term, not the famous concept of the early Foucault related, first and foremost, to the historico-cultural dimensions.

World Wars that a "turning from the world of science to the world of life" took place, and the humanistic paradigm that this turning brought with it can be called, using a well-known concept of the later Husserl, the "humanistic paradigm of the life-world." This turning continued and evolved. Heidegger occupies his own particular position with respect to this movement, but in any case he too made his contribution to it, if only with his words at the end of his "Letter on Humanism": "Future thought is no longer philosophy." In the most recent period this turning is being called an "anthropological turn," and the anthropological *epistēmē* can be regarded as the project that corresponds to the final form and logical culmination of this turn. As for the relation of this project to the "paradigm of the life-world," it can be seen as a relation of succession rather than that of opposition.

Thus, anthropology, developed as a discourse of practices of the self/practices of the Border, exhibits rather unique methodological and heuristic qualities: possibilities of its transformation into an anthropological (anthropologically funded) epistēmē for humanistic knowledge, into the "science of human sciences." This new image and status of anthropology diverges from its image and status in the classical formation of scientific knowledge not less than its new essenceless and subjectless foundations diverge from the classical European model of man. In the present day this image is undoubtedly only a "project of the next generation," a remote and radical image that decisively removes anthropology from under the aegis of "philosophical anthropology" and is oriented toward forming a certain cardinally new configuration of humanistic knowledge. But in spite of this, if one tries hard enough, one can already see a certain thread leading to this project, to this future image, in the ancient culture of the self, in the foundations of the Greek care of the self as they are represented by Foucault. He finds in these foundations an apprehension of man's task as the task of the "art of life," of creating for one's life a unique esthetic style and order. Evidently this is an anthropological task to which all human activity must be subordinated; in particular, the principle of *gnōthi seauton*, too, must be included in the care of the self thus apprehended. But doesn't this fundamental structure of the care of the self already contain the principle of the Stoics, "to direct one's gaze at oneself, and to embrace with this gaze the whole universe"? Doesn't it contain the principle that in present-day language can be defined as the "anthropologization of the humanistic/human-dimensional discourses"? Isn't a tendency toward the anthropological *epistēmē* already implanted in the depths of the conception of the ancient practices of the self developed by Foucault?

So Where Shall We Sail?

Thus, if the new anthropology really actualizes a radical epistemic project, progressing toward the status and image of an *epistēmē*, toward a science of humane sciences, then one will have the full right to say that Michel Foucault's thought was at the source of this project — the same thought that at an early stage made an equally radical declaration about the death of man. But there is no lack of consistency here. The protagonist of this new anthropology is very far from being man in the way we have conceived of the human being until now.

Bibliography

Archimandrite Sophrony Sakharov. *Videt' Boga kak On est'* (We Shall See Him as He Is). Essex, UK: Stavropegic Monastery of St. John the Baptist, 1985. [In Russian]

Binswanger, Ludwig. "The Letter to Semyon Frank of 11 November 1936." In *Semen Lyudvigovich Frank*, ed. V. N. Porus. Moscow: ROSSPEN, 2012. [In Russian]

Blanchot, Maurice. "Opyt-predel" (Limit-Experience). In *Tanatografiya Erosa* (Thanatography of Eros), ed. S. L. Fokin. St. Petersburg: Mifril, 1994. [Russian translation]

Brown, Peter. *The Body and Society: Men, Women, and Sexual Renunciation in Early Christianity*. New York: Columbia University Press, 1988.

Cadava, Eduardo, Peter Connor, and Jean-Luc Nancy, eds. *Who Comes after the Subject?* New York and London: Routledge, 1991.

Cassian, John. *Pisaniya* (Works). Moscow: Trinity Lavra of St. Sergius, 1993. [Russian translation]

Condorcet, Marie-Jean. *Eskiz istoricheskoj kartiny progressa chelovecheskogo razuma* (Essay on the Historical Picture of the Progress of Human Reason). Moscow: Sotsekgiz, 1936. [In Russian]

Deleuze, Gilles. *Foucault*. Paris: Les éditions de Minuit, 1986.

———. *Logika smysla* (The Logic of Sense). Moscow: Akademiya, 1995. [Russian translation]

———. *Nicshe i filosofiya* (Nietzsche and Philosophy). Moscow: Ad Marginem, 2003. [Russian translation]

———. *Peregovory* (Negotiations). St. Petersburg: Nauka, 2004. [Russian translation]

Bibliography

———. *Pourparler 1972-1990*. Paris: Les éditions de Minuit, 1990.
Eribon, Didier. *Michel Foucault*. Moscow: Molodaya Gvardiya, 2008. [Russian translation]
Florovsky, Georgij V. *Vostochnye Otsy IV veka* (Eastern Fathers of the 4th Century). Paris: YMCA-Press, 1931. [In Russian]
Foucault, Michel. *Abnormal*. St. Petersburg: Nauka, 2004. [Russian translation]. Original: *Les Anormaux*. Paris: Gallimard, 1999.
———. *Dits et écrits* I, 1954-1975, no. 163. Paris: Gallimard, 2001.
———. *Dits et écrits* II, 1976-1988. Paris: Gallimard, 2001.
———. *Germenevtika sub'ekta*. St. Petersburg: Nauka, 2007. [Russian translation]. Original: *L'herméneutique du sujet*. Paris: Gallimard, 2001.
———. *Histoire de la sexualité*, vol. 3: *Le souci de soi*. Paris: Seuil/Gallimard, 1984.
———. *The History of Sexuality*, vol. 1. London: Penguin Books, 1984.
———. *The History of Sexuality*, vol. 2: *The Use of Pleasure*. Harmondsworth and New York: Viking, 1985.
Gros, Frédéric. "O kurse 1982 goda" (On the Lecture Course of 1982). In Michel Foucault, *Germenevtika sub'ekta* (Hermeneutics of the Subject). St. Petersburg: Nauka, 2007. [Russian translation]
Hadot, Pierre. *Chto takoe antichnaya filosofiya?* Moscow: Izdatelstvo Gumanitarnoi Literatury, 1999. [Russian translation.] Original: *Qu'est-ce que la philosophie antique?* Paris: Gallimard, 1995.
———. *Dukhovnye uprazhneniya i antichnaya filosofiya* (Spiritual Exercises and Ancient Philosophy). Moscow-St. Petersburg: Stepnoy Veter, 2005. [Russian translation.] Original: *Exercices spirituels et philosophie antique*. Paris: nouvelle edition, 2002.
Harpham, Geoffrey. "Old Water in New Bottles: The Contemporary Prospects for the Study of Asceticism." *Semeia* 58 (1992).
Horujy, Sergej S. *Diogenes' Lantern*. Moscow: Institut filosofii, teologii i istorii Sv. Fomy, 2010. [In Russian]
———. *Issledovaniya po isikhastskoj traditsii* (Studies in the Hesychast Tradition), vols. 1 and 2. St. Petersburg: Russkaya Khristianskaya Gumanitarnaya Akademiya, 2012. [In Russian]
———. *K fenomenologii askezy* (Toward a Phenomenology of Ascesis). Moscow: Izd. Gumanitarnoj Literatury, 1998). [In Russian]
———. "Man's Three Far-away Kingdoms: Ascetic Experience as a Ground for a New Anthropology." *Philotheos. International Journal for Philosophy and Theology* 3 (2003): 53-77.
———. *O starom i novom* (Of Things Old and New). St. Petersburg: Aleteja, 2000. [In Russian]

———. *Ocherki sinergijnoj antropologii* (Essays in Synergic Anthropology). Moscow: Institut filosofii, teologii i istorii Sv. Fomy, 2005. [In Russian]

———. *Posle pereryva* (After the Interruption). St. Petersburg: Aleteja, 1994. [In Russian]

Isaac the Syrian. *Slova podvizhnicheskie* (Ascetic Orations). Moscow: Pravilo Very, 1993. [Russian translation]

Kierkegaard, Søren. *Either-Or*, vol. 2. Translated by Walter Lowrie. Princeton: Princeton University Press, 1974.

Lossky, Vladimir N. *Ocherk misticheskogo bogosloviya Vostochnoy Cerkvi* (The Mystical Theology of the Eastern Church), Bogoslovskie Trudy, vol. 8. Moscow: Izdatelstvo Moskovskoi Patriarkhii, 1971. [In Russian]

Mansurov, Fr. Sergii. *Ocherki iz istorii Cerkvi* (Sketches from the History of the Church), Bogoslovskie Trudy, vol. 7. Moscow: Izdatelstvo Moskovskoi Patriarkhii, 1971. [In Russian]

Meyendorff, John. *Vvedenie v sviatootecheskoe bogoslovie* (Introduction to Patristic Theology). New York: Religious Books for Russia, 1982. [In Russian]

Nietzsche, Friedrich. *K genealogii morali* (On the Genealogy of Morals). In F. Nietzsche, *Sochineniya v dvukh tomakh*, vol. 2. Moscow: Mysl', 1990. [Russian translation]

———. *Sumerki idolov* (Twilight of the Idols). In F. Nietzsche, *Sochineniya v dvukh tomakh*, vol. 2. Moscow: Mysl', 1990. [Russian translation]

Philokalia, vols. 1-5. Moscow: Trinity Lavra of St. Sergius, 1992. [Russian translation]

Ricoeur, Paul. *Ya-sam kak drugoi* (Oneself as Another). Moscow: Izdatelstvo Gumanitarnoi Literatury, 2008. [Russian translation]

Rossius, Andrey A. "Vvedenie" (Introduction). In Friedrich Nietzsche. *Rozhdenie tragedii* (The Birth of Tragedy). Moscow: Ad Marginem, 2001. [In Russian]

Theophan the Recluse. "Umnoe delanie o molitve Iisusovoj" (Noetic Practice Related to the Prayer of Jesus). In *Sbornik pouchenii Sv. Otcov i opytnyh ee delatelej* (Collection of Teachings of the Holy Fathers and of Experienced Practitioners of the Prayer of Jesus), ed. Igumen of the Valaam Monastery Khariton. 3rd ed. Sergiev Posad: Svyato-Troitskaya Sergieva Lavra, 1992. [In Russian]

Index

Althusser, Louis, 96
Anthony the Great, 48, 111
Anthropological Border, 123-26, 128-29, 135, 136n.17, 139-42, 151, 155n.35, 167, 171-72, 174
apatheia, 44, 111
Arendt, Hannah, 168
Aristotle, 23, 72, 73n.82, 75, 99, 115, 162, 165
ascesis (asceticism), 10, 25, 44, 57, 60, 111, 156; Christian, 2, 5, 12, 17, 26-27, 40, 46-47, 49, 50-51, 54-56, 58-59, 69, 94, 99, 101, 103, 109, 120, 122-23, 136n.17, 156-57, 159-61; homosexual, 83; philosophical (pagan), 25-26, 40-41, 43, 47-50, 136n.17, 153, 161
ascetic consciousness, 57, 106, 111-13
Athanasius the Great, 48
attention (as spiritual exercise), 43, 112-13, 158n.36
Augustine, 48, 69, 70-71, 73

Barsanuphius the Elder, 119
Basil the Great, 48
Binswanger, Ludwig, 67-68
Blanchot, Maurice, 147-48
Borderless Man, 139-42
Brown, Peter, 13, 52, 61, 163

care of the self, 6, 16-19, 23-25, 27-32, 42, 44, 47-48, 51, 72, 74, 76-78, 80, 84, 94, 133, 135, 138, 144, 172, 174
Cassian, John, 47, 49, 51, 55, 57-60, 62, 69, 70, 111, 120, 123, 156-58
Chekhov, Anton, 158
Chrysostom, John, 48
Clement of Alexandria, 48, 59
Climacus, John, 120, 157
Communion, 33, 36, 56-57, 101-2, 104, 106, 113-16, 119, 122-25, 127-28, 138, 150, 156, 159, 161, 171
confession, 3, 5, 45, 48, 52-53, 58, 61, 63-64, 70-71, 90-91, 96, 108, 122-23, 171
contemplation, 56, 94, 128
conversion to the self, 6, 12-16, 21, 32, 37-38, 62, 82, 93-94, 107-8, 110, 157, 172-73. *See also epistrophē*
Cynics (cynicism), 16, 39, 51, 80n.89

deification, 57, 104, 109, 114, 116, 131, 150, 157
Deleuze, Gilles, 6, 22-23, 89, 129, 132-35, 140, 143-48, 150-52, 155n.35, 167-70
Descartes, René (Cartesianism), 18, 20-21, 46, 73, 75-76, 99, 140, 165
dispassion, 44, 111

INDEX

Eastern Christian discourse, 100-102, 109, 115-16, 122-23, 137, 156
Epictetus, 37-38, 112, 146, 153
Epicurus (Epicureanism), 16, 37, 39, 85, 89, 118, 171
epistrophē, 37. *See also* conversion to the self
Eribon, Didier, 63, 91
esthetics of existence, 43n.35, 78, 80-88, 90-91, 95-96, 132, 143, 145, 148-49, 150, 155n.35, 169
Evagrius of Pontus, 157
exagoreusis, 53-55, 58, 119-20
exomologesis, 53-54, 123

Flaubert, Gustav, 169
Florensky, Fr. Pavel, 4n.8
Florovsky, Fr. George, 115, 162
Freud, Sigmund, 138, 142
Freudian slips, 130

Gregory of Nyssa, 48-49, 51, 69, 70-71
Gros, Frédéric, 1-2, 6n.11, 51, 66, 79, 84-86, 148-49, 154, 173

Hadot, Pierre, 12-14, 24, 37, 39-43, 51-52, 62, 64n.70, 71, 76, 78, 85-86, 91-94, 110, 149
haplosis, 128, 159
Hegel, G. W. F., 20, 74-77
Hegesias, 169-70
Heidegger, Martin, 19, 68, 74-75, 101n.4, 124, 148, 168, 174
hesychasm, 55-57, 101-2, 104, 106, 108-16, 119-21, 123-25, 127, 138, 156-59, 161, 163, 171
Hierocles the Pythagorean, 161
Hierocles the Stoic, 89
Husserl, Edmund, 68, 74-75, 174
Hypostasis, 115, 127-28, 159, 162
Hyppolite, Jean, 77, 142

identity (self-identity), 6-7, 22, 50, 54, 101, 115-16, 123, 125-31, 143, 159
Ignatius Loyola, 39-40

intentionality (intentional consciousness), 112-14
Isaac the Syrian, 111, 157

John the Elder, 119
Jung, Karl Gustav, 120

Kant, Immanuel, 73, 75, 76-77, 99, 165
Karsavin, Lev Platonovich, 116
Kierkegaard, Søren, 8-9, 77, 86, 89, 92-96, 148, 155, 159
Kojeve, Alexandre, 77

Lacan, Jacques, 74
Lossky, Vladimir, 159

Macarius of Egypt, 157
Mansurov, Fr. Sergii, 160
Marcus Aurelius, 37, 46
Marxism, 62-63, 74, 88, 165
metanoia, 37, 62, 93, 107, 161
Meyendorff, Fr. John, 49
monasticism, 25, 47, 49, 52-55, 70, 119-20

Nancy, Jean-Luc, 143n.25
neoplatonism, 14, 16, 31, 38, 56, 71, 110, 128, 159, 163
neotribalism, 88-89, 132-34, 145, 150, 170
Nietzsche, Friedrich, 22, 50, 74-75, 89, 133, 135, 137n.18, 144, 148, 170

obedience, 54-56, 58, 110
Ontic Man, 126, 128, 142
Ontological Man, 126-28, 131, 135-37, 141
organon, 103-4, 117-18, 131, 135-37, 141. *See also* spiritual tradition
Origen, 49, 162

Palamas, Gregory, 114
paraskeuē, 40-41
parrhēsia, 3, 33-37, 45, 156
passion, 12, 22, 30-31, 43n.35, 44-45, 58-59, 83-84, 91, 107, 110-11, 141, 149, 153, 155n.35, 163
Patanjali, yogi, 128
perichoresis, 115-16, 127, 138, 150, 159

Index

Philo of Alexandria, 43n.35
Plato (Platonism), 10, 12, 14-16, 19-22, 24, 27-31, 37-39, 43, 69, 72, 85, 156, 169
pleasure, 3, 13, 38, 42, 69, 80, 83-84, 87, 89-90, 100, 132, 134, 149, 152, 155n.35, 168-69, 170
power, 1-2, 10, 31-32, 49, 52, 64, 168
prayer, 113, 116, 157-58
psychagogy, 12, 32-33
psychoanalysis, 74, 82, 86, 110n.7, 129, 138, 170n.5
purification, 6, 28, 72, 110
Pythagoras (Pythagorean, Pythagorism), 25, 28-29, 32, 161

renunciation (self-renunciation), 6, 27, 29, 48-52, 54, 60, 65, 69, 71, 94-95, 108-10, 158-59
repentance, 5, 48-49, 53-54, 62-63, 96, 107-10, 123, 157-58, 160-61
Ricoeur, Paul, 20n.19

scholasticism, 72, 73n.82, 75
self: ancient practices of the, 21, 80-81, 114, 117, 121, 138, 163, 166, 174; possessive, 117, 137; true, 7-9, 14, 23-24, 30, 32, 34-35, 38, 93, 96-97, 110
Seneca, 37-39, 45-46, 91-92, 146, 153, 173
shamanism, 136-37
sobriety, 112
Socrates, 19, 28, 31, 42, 70, 92, 120, 122
Source-beyond-there, 104, 106, 108-9, 114, 129, 134n.15, 156

spiritual exercises, 12, 33, 35, 39-45, 48, 51, 64, 76, 78, 92, 111, 114, 136, 148-49, 154
spiritual tradition, 36-37, 117-21, 137, 151, 156-57, 163, 171. *See also organon*
Stoics (Stoicism), 16, 38-40, 43, 46, 51-53, 78, 80n.89, 85, 89, 91-92, 96-97, 111, 114, 121, 149, 153, 158, 161, 163, 172-74
subject: constitution of the, 8, 11, 26, 47, 64, 66, 70, 91-93, 95-96, 100, 102; first person, 117, 137-38, 156
subjectivation (self- and trans-subjectivation), 14, 16, 22-23, 35, 44, 53, 60, 134, 145, 153, 167-69
subjectness, 30, 41, 67, 69, 70-71, 86, 91-92, 109, 116-17, 138, 148, 155n.35, 156, 166, 169-70
suicide, 134, 139, 147n.29, 168-70
synergy *(synergia)*, 105, 114, 116, 131, 156-57

Tertullian, 48-49, 54, 69, 160
Theodor the Studite, 119
Theophan the Recluse, 113, 128
theosis, 57, 104, 109, 114, 116, 131, 150, 157
truth: access to, 6-8, 18, 34-36, 72-73, 75; game of, 65, 67; truthfulness, 35, 37

unconscious, 124, 126, 128-31, 136-42, 149, 155
unlocking, 8, 86, 105, 110, 114, 116-17, 124-26, 131, 133-35, 137-38, 140, 154, 156, 166

Virtual Man, 124-26, 130-31, 135, 142-43, 146, 151, 158, 160